Ecology of plant communities

Ecology of plant communities

*A phytosociological account of
the British vegetation*

JACK RIELEY

Department of Botany, University of Nottingham, UK.

SUSAN PAGE

Department of Adult Education, University of Leicester, UK.

Longman
Scientific &
Technical

Copublished in the United States with
John Wiley & Sons, Inc., New York

Longman Scientific & Technical
Longman Group UK Limited
Longman House, Burnt Mill, Harlow
Essex CM20 2JE, England
and Associated Companies throughout the world.

Copublished in the United States with
John Wiley & Sons Inc., 605 Third Avenue, New York,
NY 10158

© Longman Group UK Limited 1990

First published in 1990

British Library Cataloguing in Publication Data
Rieley, J. O.
  Ecology of plant communities.
  1. Great Britain. Plants. Ecology
  I. Title   II. Page, Susan
  581,5′0941

ISBN 0-582-44639-2

Library of Congress Cataloguing-in-Publication Data
Rieley, J. O., 1941–
  Ecology of plant communities.

  Includes bibliographical references.
  1. Plant communities – Great Britain. 2. Botany –
Great Britain – Ecology. 3. Phytogeography – Great
Britain. I. Page, Susan, 1957–   . II. Title.
QK306.R54 1990   581.5′247′0941   89–12944
ISBN 0-470-21482-1

Set in Linotron 202 9/12 pt Times

Produced by Longman Singapore Publishers (Pte) Ltd.
Printed in Singapore

Contents

Preface

Over many years of teaching plant ecology to under-graduate and adult education students it became obvious to us that there was no book suitable for our courses that described adequately the plant communities of the British Isles. *Ecology of Plant Communities* was written to go some way towards filling this gap, and is intended to provide a reference text on the principal British vegetation types for university, poly-technic and college students. It will be particularly useful to those studying or teaching courses in botany, biology, ecology, environmental studies, geography and allied topics, including field studies, and to anyone with a professional interest in nature conservation and vegetation management.

The phytosociological approach to plant community classification has been adopted by us throughout and the associated nomenclature to indicate the various levels in the hierarchical classification is used wherever possible. These methods have been practised success-fully in most of continental Europe for over 60 years and are being increasingly applied to the British vegetation. It is hoped, therefore, that this book will be of particular value to the growing number of British botanists, ecologists and plant geographers who recognise this method of vegetation description, as well as to the large body of European plant ecologists who will welcome the integration of the British vegetation into a common system. Also, on reading this book members of the international community of botanists and plant geographers will gain an insight into the vegetation of the British Isles, the major factors influencing the development of our plant communities and the relationship of British vegetation to the major plant formations of the world.

Since it is interaction between various factors that determines the evolution and succession of plant communities it is logical to commence the book and individual chapters with information on the environ-mental influences that control plant community development and geographical distribution. Prominence is also given, in appropriate chapters, to the importance of historical changes and anthropogenic factors. These introductory sections are followed by detailed descriptions of the principal plant communities of woodland, grassland, heath, freshwater wetland, peatland, salt marsh, sand dune, mountain and urban ecosystems.

The abiotic factors of light, temperature, wind, water and nutrient supply, and the biotic pressures of migration and competition which collectively determine plant community composition are presented in Chapter 1. The environmental influences which operate through the soil are considered in Chapter 2, which also provides information on pedogenesis and the principal soil types of the British Isles. In Chapter 3 the techniques of phytosociology, the classification of vegetation based upon floristics, are described and a synopsis of the higher phytosociological classificatory units used in the text is provided. The highest phytosociological unit is the vegetation circle or plant formation and an overview of these is presented in Chapter 4 to set the scene for the account of the plant communities of the British Isles that follows in subsequent chapters.

Chapter 5 commences with a brief historical review of British woodlands since the end of the last glaciation, followed by an account of the plant communities of woodlands, lowland grasslands and dry heaths. In Chapter 6 the many facets of water and its influence on plant community structure are detailed emphasising the differences between communities of deep and shallow, eutrophic and oligotrophic, slow and fast flowing waters. The conditions necessary for the development and maintenance of acid peat bogs are explained in Chapter 7 together with information on the accompanying hydrological and hydrochemical changes that take place as mire succession proceeds. Consideration is given in Chapter 8 to the coastal ecosystems of sand dune and salt marsh. The effects of salinity and substrate instability on plant survival and community structure are discussed and the various plant communities are described with reference to the vegetational zonations which · occur in these coastal habitats.

Increases in altitude and latitude bring about changes in all of the primary environmental factors which interact to delimit plant communities. These modifications and their implications for the vegetation of montane and northern regions of the British Isles are detailed in Chapter 9. Finally, the special features of the urban ecosystem are assessed in Chapter 10. Habitat diversity may be greater in cities than in the surrounding countryside owing to the diversity of semi-natural and man-made habitats the flora of which is enhanced by the presence of adventive and alien species.

Inevitably in writing a book of such wide ranging scope a few plant communities of localised distribution or specialised habitats have been omitted but otherwise these chapters provide a comprehensive account of the ecology of the British vegetation.

Assistance and encouragement to the authors whilst they were writing this book has been provided from many sources and a few warrant special mention: Messrs Longmans for their forbearance over several years and to the series editor Professor Dennis Baker for inviting us to write the book in the first place and remaining understanding through numerous delays; generations of students who have been the testing ground for many of the ideas and material that are contained herein; Brian Case and Peter Smithurst of the Department of Botany, University of Nottingham for photographic assistance and technical expertise, respectively; Peter Shepherd for information on and photographs of urban habitats; Dr. Andrew Lister for photographs of Antarctica; and last, but certainly not least, to our families and friends who provided us with the incentive to keep going.

We have been fortunate in that almost all of the photographs used in this book were taken by ourselves and this has avoided the inconvenience of acquisition and copyright. Figures and tables for which authorisation has been obtained are gratefully acknowledged.

The sources used for plant nomenclature are (a) for higher plants – Clapham, Tutin & Moore (1987); (b) bryophytes – Corley & Hill (1981); (c) common names of higher plants – Dony, Perring and Rob (1980).

J. O. R.
S. E. P.

March 1989

Acknowledgements

We are grateful to the following for permission to reproduce copyright material:

Academic Press Inc. and the author, L. F. Curtis for fig. 2.2 from fig. 1, p. 59 (Cruise & Newman, 1973); Edward Arnold for fig. 4.2 (Walter, 1973) and fig. 9.1 from fig. 2.17 (Dajoz, 1969); Benjamin/Cummings Publishing Co. for fig. 1.5 from fig. 10.5, p. 205 (Barbour *et al.*, 1987); Blackwell Scientific Publications Ltd. for fig. 7.4 from fig. 6, p. 245 (Bartley, 1960) and table 7.3 from table 17, p. 345 (Crisp, 1966); Botanical Society of the British Isles for figs. 9.5b, 9.6b and 9.7b from pp. 138, 75 and 139 (Perring & Walters, 1962); Cambridge University Press for fig. 2.5 from figs. 27–30, pp. 87–93 (Tansley, 1949) and table 6.1 from p. 201 (Ratcliffe, 1977); Hodder & Stoughton Ltd. for fig. 5.1 from fig. 1.1, p. 4 (Moore & Webb, 1978); Macmillan Publishing Co. for fig. 4.1 from fig. 3.8, p. 65 (Whittaker, 1975), copyright © 1975 by Robert H. Whittaker; the author, Dr. P. Moore for fig. 2.7 from fig. 2.3 (Moore & Bellamy, 1973); Redaction Candollea-Boissiera for fig. 4.9 from fig. 2, p. 573 (Bourgeron & Guillaumet, 1982); Verlag Eugen Ulmer for fig. 6.3 from fig. 205, p. 364 (Ellenberg, 1963); Westfälisches Landesmuseum für Naturkunde for table 10.1 from table 1, p. 36 (Horbert, 1978); John Wiley & Sons Ltd. for fig. 2.6 from fig. 3.10, p. 92 and table 1.3 from p. 280 (Etherington, 1975).

We are unable to trace the copyright owner of fig. 9.5a which appeared as fig. 27, p. 36 of *The Circumpolar Plants, II. Dicotyledons* by E. Hulten, published by Almqvist & Wiksell (1971) and would appreciate any information which would enable us to do so.

Chapter 1

Introduction: factors influencing plant distribution

The geographical distribution of plant species is to a very large extent determined by variation in environmental conditions. A large number of interrelated factors simultaneously affect the growth, development and reproduction of organisms; they operate together as a single dynamic system so that it is impossible to predict the effect of one factor in isolation from all others. However, despite this interplay, a general distinction is usually made between the abiotic (non-living) and the biotic (living) aspects of the environment (although this separation is not strictly justified). Examples of abiotic environmental factors which affect plants are temperature, light (quality and quantity), soil structure, fire, and water and nutrient supply (section 1.1). Biotic features include genetic potential and variation within the species, plant/animal interactions of pollination and seed dispersal, and plant/plant inter-relationships of competition, mutualism and parasitism (section 1.2). (Odum, 1971). Directional, non-random change in vegetation occupying a given area, through time, may also take place. This succession (section 1.3) of communities may be induced by an environmental change or by the intrinsic properties of the plants. (Kershaw, 1973).

Natural plant populations are genetically diverse and through the processes of evolution and natural selection, the 'fittest' individuals, i.e. those most suited to the environmental conditions under which they live, will survive to reproduce, thus maintaining the population. This 'survival of the fittest' is a response to the abiotic and biotic features of the environment – through selection of those genotypes which impart preferential 'adaptations' on members of a population. As a consequence of evolution and selection processes each plant species has an inherited ability to survive within a limited range of environmental conditions known as its tolerance range or ecological amplitude. The greater the genotypic variation expressed by a population, the greater its ecological amplitude is likely to be.

1.1   Abiotic factors

The principal abiotic factors which influence plant growth and distribution are those of the atmosphere (climatic factors) and of the soil (edaphic factors) (Ch. 2). Climate is usually described as a variety of separate components – rainfall, temperature, wind and light which pertain over a whole region. This is the 'macroclimate'. Small, localised variations arising from differences in the nature and extent of vegetation cover, soil conditions and microtopography are responsible for the 'microclimate'.

1.1.1.   *Light*

Light is a form of physical energy which includes not only the visible part of the spectrum but also ultraviolet, infra-red and other invisible radiations. The quality and quantity of light affects plant growth and development. Visible light is essential to all green plants as the energy source for photosynthesis; infra-red and far-blue radiations play an important role in photoperiodism; ultraviolet radiations, however, can be harmful to living organisms (section 9.2.3).

Latitudinal position affects the amount of solar radiation reaching the earth's surface: at midday on the equator the sun's energy penetrates the atmosphere at near enough 90°, whilst at latitudes north or south the angle becomes increasingly oblique and solar radiation is reduced. This is further altered in hilly areas, depending on aspect (i.e. in the northern hemisphere north-facing slopes receive considerably less solar radiation than south-facing slopes and vice versa in the southern hemisphere). The geographical distribution of global plant formations is greatly influenced by changes in irradiance with latitude (Ch. 4). The light requirements of different plant species result in the development of complex stratified communities.

Green plants vary in their requirement for light and many species have morphological and physiological adaptations which maximise their growth efficiency under different light intensities. (Larcher, 1975; Street & Opik, 1984). Arable weeds (for example, *Achillea millefolium* (yarrow)) which flourish in open areas under full light intensity are heliophytes (sun-adapted

plants), which possess small, much-divided leaves, and are often short-lived or ephemeral. They depend on the rapid production of large numbers of seeds for survival. In contrast, many woodland understorey plants (e.g. *Mercurialis perennis* (dog's mercury)) are sciophytes (shade-adapted plants), the leaves of which are characteristically thin with a large surface area to maximise photosynthesis under the low-light conditions in which they grow. Sciophytes usually have a long vegetative growth period, a reduced dependence on flowering and seed production, and well-developed strategies for vegetative reproduction by bulbs, corms, rhizomes or tubers.

Light intensity is also an important factor determining plant growth in aquatic environments (section 6.5).

1.1.1.1   *Photoperiodism*

The duration of periods of light and dark, whether daily or seasonal, acts as a trigger mechanism for many plant responses, including initiation of flowering and dormancy, time of leaf fall, initiation of growth of tubers and bulbs, development of pigmentation and regulation of the supply of mineral nutrients to various organs within the plant. Photoperiodic responses ensure that plant growth coincides with the most favourable season; photoperiodic control of flowering also ensures that bird- and insect-pollinated plants produce their flowers at the time when suitable pollinators are available. (Vince-Prue, 1975).

In woodlands, the annual sequence of vegetative growth and flowering is strongly correlated with the annual light curve underneath the tree canopy. In temperate regions, many spring and early summer flowering species respond to the shorter daylength of these seasons and carry out the important stages of their reproductive cycle before the tree canopy is fully developed and the woodland floor is shaded. (Tansley, 1939).

1.1.2   *Temperature*

With the exception of a few specialised bacteria and algae which can tolerate the extreme cold of polar and high alpine regions, or the near-boiling temperatures of hot springs, most living organisms can survive only within a relatively narrow temperature range. The most obvious effect of a change in temperature is on metabolism. With an increase in temperature both photosynthesis and respiration increase, up to around 35°C for the former and 45°C for the latter (provided that no other factors are limiting). At higher temperatures most enzymes are destroyed.

Temperature is broadly influenced by latitude and the range of temperature which a plant can tolerate is a primary factor influencing the geographical distribution of plant formations (Ch. 4). At a given level of solar radiation each species has an optimum temperature at which germination, seedling establishment and growth will be maximal. However, it is often difficult to separate direct temperature effects from those induced by other environmental factors, in particular the availability of water. This is of particular relevance in tropical regions, where high air temperatures lead to increased transpiration from plant tissues. In regions of low or irregular rainfall, the combination of high temperatures and low moisture availability can impose severe drought stress on the vegetation (section 1.1.3.1).

The most widespread effect of low temperature upon plants is frost injury. Although tropical plants may be killed at temperatures above freezing, those of colder climates are adapted to withstand very low temperatures. In regions where frosts are frequent many plants become 'acclimatised' to the cold (section 9.2.3). Diurnal temperature fluctuations can be extreme in some environments, e.g. deserts and high mountains (sections 4.8 and 4.9), but variations are reduced under vegetation canopies. In forests, for example, daytime temperatures are lower under the tree canopy than outside because of the shade and the higher relative humidity; overnight, temperatures fall at a lower rate under vegetation canopies than outside.

Additional environmental influences on temperature include altitude (air temperatures fall with increasing elevation – section 9.1.2), aspect (the direction of a slope affects the amount of solar radiation it receives), proximity to water (large water bodies, i.e. lakes and oceans, have a moderating effect on the temperature of adjacent land areas, reducing both diurnal and seasonal fluctuations) and cloud cover.

1.1.3 *Water*

All living organisms require water for their survival. The availability and abundance of this resource greatly influences the distribution of aquatic and terrestrial plants and, on a global scale, the geographical distribution of plant formations (Ch. 4). An inadequate water supply limits metabolic processes and induces wilting. Water supply to plants is dependent on the amount of precipitation, the relative humidity of the atmosphere, the level of the water table relative to the soil surface

and the rate of water loss from both the ground (evaporation) and vegetation (transpiration).

The amount of rain falling in any location is influenced by major geographical features, in particular world-wide weather systems which superimpose a dominant effect on local climatic conditions. In general, the quantity of rainfall increases with altitude up to about 1500 m in temperate, medium-latitude zones and decreases with latitude from the equator to polar regions where the low temperatures greatly reduce the water-holding capacity of the atmosphere. In addition, topographical features, distance inland from major oceans, vegetation cover and human activities superimpose variations on this scheme.

Although high rainfall is a predominant feature of the British climate it is not uniformly distributed throughout the country – ranging from 600 mm per annum in the

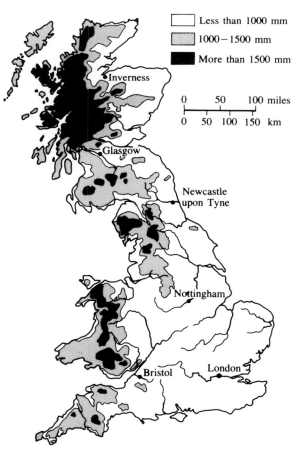

1.1 Distribution of rainfall in Great Britain (modified from the Rainfall Atlas of the British Isles, 1926).

south and east up to 5000 mm per annum in the north and west (Fig. 1.1). The quantity of precipitation, which also differs between seasons of the year, exerts a major influence on soil formation processes (section 2.3).

Plant cover has a profound effect on microclimate by intercepting rainfall and preventing direct and rapid evaporation of water from the ground (Table 1.1).

Table 1.1 Interception of rainfall by the canopy and under-canopy in Coed Cymerau National Nature Reserve, north Wales in a 20-week period (Rieley *et al*, 1979)

	Mean volume of water ($dm^3 \, m^{-2}$)	Percentage interception by each plant layer of the water reaching it	Percentage interception by the entire vegetation
Rainfall (incident)	544±1.5		
Throughfall (under tree canopy)	453±9.3	16.7	
Leachate (under bryophytes)	303±9.0	33.1	44.3

Underneath vegetation canopies the relative humidity is higher than in the open. The growth and reproduction of bryophytes, fungi and pteridophytes are dependent on high relative humidities throughout the year. These lower plants are more abundant in high rainfall regions of the British Isles or in sheltered habitats where evaporation is reduced (e.g. valleys, ravines, streamsides and rock crevices) than in low rainfall areas or exposed situations. In the open, the higher temperatures encourage continued water loss with the result that much of the incident rainfall is returned almost immediately to the atmosphere. In hot climates this leads to water deficit stress. At high latitudes the low temperatures depress the rate of evaporation and this combined with the lower rainfall of these regions may paradoxically produce high humidities close to the ground.

Water may account for up to 90% of the fresh weight of plant material and although only a small proportion of this is required for essential metabolic processes (some estimates are as low as 1%), a readily available supply in the soil is essential. Many plants, however, grow in regions of considerable water stress as a result of an inadequate supply or excessively high rates of evaporation and transpiration. Alternatively, plants may be subject to an excess of water through waterlogging of the rooting substrate. These extremes of water supply impose considerable stresses on plant growth and only

those species which have evolved some means of tolerance to these conditions (represented in appropriate morphological, anatomical or physiological adaptations) are able to survive.

1.1.3.1 *Water deficit*

Water deficiency has a direct effect on plant metabolism. In photosynthesis water shortage affects carbon dioxide supply, the photochemical reactions which combine water and carbon dioxide, the chemical reduction of carbon dioxide and the translocation of photosynthates. These limitations are brought about partly by a lack of water for the biochemical reactions to function or for translocation to occur, and partly by the closure of the stomata as the leaves wilt, which reduces, or stops, the gaseous exchange of carbon dioxide between the atmosphere and the internal leaf-cell surfaces.

The presence of xeromorphic features enables plants to avoid or postpone the more serious effects of low water availability. These attributes include: thick, waxy cuticles; stomata protected by hairs or located in chambers recessed within the surface of the leaf; stomata restricted to the lower surface of the leaf which may also be inrolled (Fig. 1.2); hairs on the surfaces of leaves which reflect radiation from the sun or rock surfaces, thereby reducing leaf temperatures; and positioning of the chlorophyllose cells in the leaves of some species underneath a layer of non-photosynthetic 'hypodermal' cells which enable photosynthesis to continue even after loss of water from the outer cells. Many xerophytes, growing in areas prone to drought, also have abundant sclerenchymatous supportive tissue which prevents permanent injury to the plants during temporary wilting. Modifications to root systems are also apparent: desert species have two types of roots – a spreading, much branched system of fine roots near to the soil surface which intercepts water falling during intermittent rain showers, and/or long, far-penetrating roots which absorb water from a deeply situated water table.

Sudden exposure of plants to water stress is more harmful than gradual exposure to increasing water deficit over a long period of time. Some plants can become 'hardened' to survive periods of drought without permanent injury. In arid regions, where high temperatures and low rainfall during the growing season are particularly stressful, many plants have developed a more efficient use of their available water supply than have species of temperate zones. These plants either

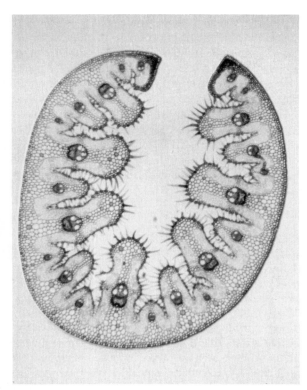

1.2 T.S. leaf of *Ammophila arenaria* (marram grass) which has xeromorphic adaptations to minimise transpiration water loss: inrolled leaf, hairs and stomata on adaxial surface, waxy cuticle on abaxial surface.

exhibit the C_4 (Hatch/Slack) pathway of carbon metabolism or Crassulacean acid metabolism (CAM). Both of these groups of plants transpire less water per gram of dry weight produced than do their C_3 counterparts with the Calvin photosynthetic pathway.

1.1.3.2 *Water excess*

Excess of water is as inimical to the growth of many plants as water deficiency. Waterlogged soils develop in regions of high rainfall and/or where drainage is impeded, either by the presence of impervious rocks or by the development of an iron pan (section 2.4.1.2) in the lower soil horizons. The main problems for plants growing in waterlogged soils are associated with chemical toxicity caused by changes in the oxygen content and pH of the substrate.

As oxygen is depleted in waterlogged soils the carbon dioxide concentration increases. In the absence of an adequate oxygen supply, some soil micro-organisms utilise iron, nitrogen, phosphorus or sulphur compounds as alternative electron acceptors to oxygen in their respiratory metabolism. Reducing conditions develop and potentially toxic substances, e.g. aluminium, iron, manganese, sulphur and phosphorus, accumulate. (Fitter & Hay, 1981; Crawford, 1983). Losses of nitrogen by denitrification can be rapid and may have serious effects on the growth of plants.

Anaerobic respiration in roots leads to the production of ethanol which, if it accumulates, has serious or fatal consequences. Some plants of wet soils (e.g. *Juncus effusus* (soft rush) and *Glyceria maxima* (reed sweetgrass)) produce metabolites other than ethanol, such as malic and shikimic acids, neither of which is toxic to plant tissues. These substances may accumulate in root cell vacuoles until aerobic edaphic conditions return, or they can be transported to shoots where they take part in aerobic respiration. In flood-tolerant species ethanol may diffuse from the roots into the air via lenticels (e.g. in *Alnus* (alder), *Betula* (birch) and *Salix* spp. (willow)), which are particularly well developed around the base of the stems. (Crawford, 1972, 1978, 1983).

Some plants of waterlogged soils release oxygen from their roots into the rhizosphere which surrounds them. This creates an oxidising micro-environment in which potentially toxic elements are reconverted to their less soluble or less toxic forms. This radial oxygen loss from roots is responsible for the precipitation of insoluble iron hydroxides on to the external root surfaces of *Molinia caerulea* (purple moor-grass) in waterlogged soils (Armstrong, 1967).

The most common anatomical adaptation of hydrophytes (plants of open water or waterlogged soils) is the development of extensive, internal air space tissue (aerenchyma) (Fig. 1.3). This provides a system of continuous air channels from shoots to roots through which oxygen moves to maintain aerobic respiratory oxidation processes, or to aid in oxygen loss from the roots. However, aerenchyma is not a structural feature of all flood-tolerant species – for example, *Phalaris arundinacea* (reed canary-grass) and *Filipendula ulmaria* (meadowsweet) do not have this tissue.

1.1.4 *Carbon dioxide supply*

Plants require a constant supply of carbon dioxide to sustain photosynthesis. Terrestrial plants utilise the carbon dioxide of the atmosphere, which averages about 0.03% at the earth's surface. An increase in the carbon dioxide concentration of the atmosphere external to a

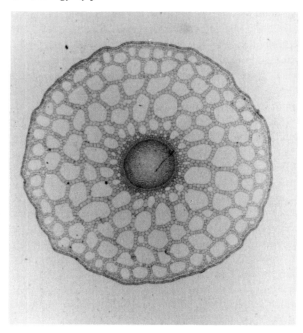

1.3   T.S. stem of *Hippuris vulgaris* (mare's tail) showing aerenchyma, an adaptation to facilitate gaseous movements between shoot and root in some hydrophytes.

plant increases the rate of photosynthesis until some other factor becomes limiting. At high altitudes the reduction in the partial pressure of carbon dioxide may limit the rate of photosynthesis (section 9.1.5).

1.1.5   *Wind*

The main influence of wind on plants is to increase the rate of transpirational water loss from the leaves. If the water supply to plant roots is insufficient to compensate, then wilting and ultimately death will ensue. Distortion of trees as a result of strong and persistent prevailing

winds is a common phenomenon and 'wind-burn' of vegetation is frequent in exposed situations, near sea coasts and at high altitudes and latitudes. In these regions tall trees cannot survive and many of the plants have a low-growing or prostrate habit and are xeromorphic. The mutual protection provided by individual species growing closely together in mats or cushions ensures their continued survival under unfavourable conditions (section 9.2.1)

Wind is also responsible for the development and maintenance of the sand dune ecosystem (section 8.3).

1.1.6   *Nutrient supply and plant nutrition*

Plants require a large number of chemical elements for their successful growth and reproduction (Table 1.2). Apart from carbon, hydrogen and oxygen, plants also need substantial amounts of calcium, iron, magnesium, nitrogen, phosphorus, potassium and sulphur. Of these 10 macronutrient elements, the first three are usually obtained from the atmosphere and soil water, whilst the others are absorbed as ions in solution in water. In addition, very small quantities of several micronutrient elements are essential for healthy plant growth, including boron, cobalt, copper, manganese, molybdenum and zinc, adequate amounts of which are available in most soils and fresh waters (Epstein, 1972; Sutcliffe & Baker, 1974). Many of these elements, e.g. copper, molybdenum and zinc, are injurious to plants if present in more than trace amounts, and certain non-essential elements, such as arsenic, cadmium, lead and mercury are phytotoxic even at very low concentrations.

Terrestrial plants obtain most of their mineral nutrients from the soil solution by diffusion, or from the soil colloid complex by cation exchange (section 2.2.4). Plant growth is limited by the low availability of one or more nutrient elements, some of which are absorbed

Table 1.2 Nutrient elements essential for plant growth. Carbon, hydrogen and oxygen come mostly from air and water; soil solids are the source of all other nutrients.

Macroelements		Microelements		Other elements which, although not universally essential may increase the growth of some plants
Carbon	Nitrogen	Iron	Copper	Sodium
Hydrogen	Phosphorus	Manganese	Zinc	Silicon
Oxygen	Potassium	Boron	Chlorine	Strontium
	Calcium	Molybdenum	Cobalt	Barium
	Magnesium			Aluminium
	Sulphur			Selenium

and retained against steep concentration gradients be-tween substrate and plant tissue. The soluble nature of most of the elements essential to plant growth means that they are readily leached out of soils in regions where rainfall exceeds evaporation. The availability (i.e. solubility) of nutrients in aquatic and terrestrial systems is also influenced by the concentration of hydrogen ions (section 1.1.6.1). Other edaphic factors which may in-fluence plant performance include soil structure, texture and cation exchange capacity (Ch. 2).

1.1.6.1 *Hydrogen ion concentration (pH)*

The pH of most mineral soils lies between 3.5 and 10.0, although peat soils may be less than 3.0 and alkaline or saline soils as high as 11.0. The pH of most freshwaters lies between 6.0 and 7.5, but that of some dystrophic peat pools and streams can be as low as 4.0, whilst calcareous or saline lakes may be as high as 11.0. Acid substrates support relatively few species whereas cal-careous ones usually exhibit considerable plant diversity.

The principal effect of soil acidity on plants is con-cerned with the uptake of nutrients by roots. This can be influenced in two ways: (a) through hydrogen ions competing for ion exchange sites; (b) indirectly, by influencing the availability of certain of the principal plant nutrient elements and also the presence of toxic ions. The availability of the most important macro- and micronutrient elements in soils of different pH values together with the relative abundance of soil micro-organisms is indicated in Fig. 1.4. As pH increases the amount of exchangeable calcium, phosphorus, boron, magnesium and molybdenum in the soil increases, whilst the availability of iron, manganese and zinc decreases. However, at pH values less than 5, iron, manganese and aluminium levels may be toxic to plants, especially when the soil is waterlogged and poorly aerated for part of the year. Best overall nutrient availability occurs in soils with a pH of about 6.5.

Bacterial and fungal activity in soils is also affected by pH. Activity is at a maximum between pH 5.0 and 8.0 and throughout this range decomposition of organic matter proceeds rapidly. Outside this pH range both bacterial and fungal decay is reduced, fewer fungal species occur and the soil fauna becomes less diverse. Reduction in decomposer activity at low pH, especially when coupled with waterlogging, poor aeration and calcium deficiency, leads to the accumulation of a layer of largely undecomposed organic material which, when consolidated, is termed peat (section 2.4.2.3).

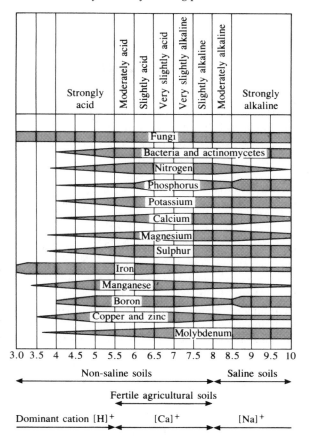

1.4 The relationship between the pH of a soil, activity of micro-organisms and availability of plant nutrients. (The width of the band represents the proportion of each element available from the maximum in the soil pool; for micro-organisms the band width represents the activity as a proportion of the optimum.)

1.1.7 *Fire*

Fire is an important ecological factor in both forest and grassland plant formations of temperate zones, and in the dry tropics (sections 4.4 and 4.6). Fires can arise naturally from lightning or spontaneous combustion, particularly in hot, arid regions, or be induced by man as part of land clearance strategies prior to cultivation, afforestation or other forms of management regime (e.g. heather moor management (section 5.4)).

In some ecosystems, particularly temperate and tropical grasslands, occasional fires maintain the tree-less, open character of the vegetation, although destruc-tion of the surface vegetation may be followed by soil instability and erosion. In habitats where fire is a frequent phenomenon, fire-resistant or even fire-dependent plants may dominate. For example, the seeds of *Pinus divaricata* (jack pine) are only released from the pine cone after it has been heated to a high enough temperature. In some parts of the south-eastern United States where forest fires are of regular occurrence, *Pinus divaricata* has replaced other less resistant *Pinus* spp. as the dominant forest tree. (Pears, 1977).

1.2   Biotic factors

The direct effect of one organism on another may be treated as an environmental factor. The interactions between biotic agents (micro-organisms, animals, other plants) may be highly specific (e.g. the host/parasite relationship) or of a more general nature (e.g. defoliation of vegetation by herbivorous animals). In contrast to abiotic environments, their biotic counterparts tend to be highly dynamic; as Harper (1977) writes, 'in an environment dominated by biotic forces . . . the evolutionary game is potentially unending. Organisms form part of each other's environment; a change in the nature of any one component changes the environment and so changes the selective forces acting on others and there is further feedback in turn as they change.'

Biotic interactions may be both beneficial and inhibitory; they include the ability of plants to maintain and extend their distributions by dispersal and migration; competition between plants for limited resources; and mutually beneficial relationships between plants and micro-organisms.

1.2.1   *Disperal and migration*

Dispersal is the movement of individuals of a species into and out of habitats and can be achieved by both vegetative and sexual reproductive agencies. In some habitats dispersal provides many ecologically diverse species as colonisers, whilst in others, because of geographical or ecological barriers, only a few specialised colonisers may be available. The ability of species to migrate into areas in which they were not formerly present and to increase their distribution range by dispersal is of paramount importance to their survival. Plants have evolved many different dispersal methods, involving various agents (wind, water and animals) for distributing their reproductive propagules (seeds and spores) to new sites.

1.2.2   *Species-species interactions*

Interrelationships between organisms are of several types: some interactions are negative, as in competition, in which one organism is inhibited in favour of the other; others are positive, in which case one or other organism is stimulated as in commensalism and mutualism.

1.2.2.1   *Mutualism*
In interactions of this type species operate to their mutual advantage; in the absence of the interaction the growth of both species is depressed. Examples of mutualistic associations are lichens (alga and fungus); mycorrhizae (fungus and higher plant); symbiotic nitrogen fixation (bacterium or blue-green alga and higher plant); and pollination (insect, bird or mammal and flowering plant). Mutualistic or symbiotic relationships are important to many aspects of the life histories of organisms since they are commonly associated with either obtaining food or, in animals, avoiding predation.

1.2.2.2   *Commensalism*
In this relationship one of the partners derives some positive advantage while the other apparently does not. True epiphytes, for example, do not obtain their food from the host upon which they grow but simply derive physical anchorage and position. Epiphytes include herbaceous perennials, ferns, mosses, algae and lichens.

1.2.2.3   *Competition*
In the absence of restraints populations expand to fill and eventually overfill the space available to them. A major control of population numbers is through competition, which occurs when the use of a resource – food, water, space or light – by one individual reduces the amount available to another individual of the same or a different species (intraspecific or interspecific competition, respectively). (Begon *et al.*, 1986).

The activities of competing individuals reduce a shared resource to the extent that interactions become mutually detrimental. The result of competition at the population level is that density (rate of population energy flow) is reduced or retarded by the competitive action. The factors for which plants compete, the characteristics of species which lead to competition and the external factors which influence competition are summarized in Table 1.3.

Table 1.3 Analysis of plant competition in terms of the factors for which plants compete or which are related to competition between plants (from Etherington, 1975).

A. Factors for which plants compete	(a) Space	Above and below ground
	(b) Light	
	(c) Carbon dioxide	} Above ground
	(d) Nutrients	
	(e) Water	} Below ground
B. Plant characteristics which cause competition	(a) Passive root interactions such as the normal production of respiratory CO_2	
	(b) Direct interaction due to the secretion of specific toxins into the environment (allelopathy)	
C. Interactions with external factors which influence or cause competition	(a) Competition for pollinators	
	(b) Competition for agents of dispersal	
	(c) Selective pressure or disturbance of ecological equilibria by animals and man	
	(d) Disturbance of the environment which provides bare soil or seedling niches	
	(e) Influence of temperature, humidity, exposure, wind, etc. on other competitive factors	
	(f) Unusual soil conditions such as toxic solute content, heavy metals, excess calcium carbonate, etc.	

1.3 Succession

Succession is the dynamic process by which populations of plants and animals replace one another over a period of time until relative community stability is achieved. The series of successional communities is termed a 'sere' (Latin *sertus* – joined or connected). The early seral stages are formed by pioneer communities, which encounter extreme habitat conditions, especially of moisture, temperature and light intensity. As environmental conditions change and the habitat is altered, for example by shading or incorporation of organic matter into the soil, pioneer species are replaced by longer-lived species with different habitat preferences. The end-stage is the climax community, which is in a state of equilibrium with prevailing environmental conditions. Some doubt has been expressed, however, as to whether stable climax communities can actually be recognised in the field. A more pragmatic approach is to identify communities in which the rate of change of succession has slowed to the point where any change is imperceptible. (Begon *et al.*, 1986).

Successions may be either progressive (e.g. bare ground to forest) or regressive (e.g. eroding peat-covered moorland). In the former, the considerable initial increase in species diversity eventually gives way to a complex structural diversity. Retrogressive succession takes place when the ecosystem becomes degraded in some way, for example, from loss of topsoil or leaching of soil nutrients, and species diversity declines. Successions can also be classified as primary or secondary, depending on whether or not the substrate on which ecosystem development commences has previously supported any vegetation. Primary successions originating on dry, terrestrial surfaces are known as xeroseres (examples include the psammosere initiated on bare sand (section 8.3)), in open freshwater as hydroseres (section 6.4) and in brackish water as haloseres (section 8.2). Primary successions commencing with plant colonisation of bare ground are rare events since these follow dramatic disturbance to or upheaval of land masses, e.g. volcanic activity, lowering of water levels or glacial movements. As a result of human interference with vegetation, examples of primary successions are less commonly encountered than secondary successions.

Secondary successions occur on sites which, although bare, have in the past supported vegetation. Removal of plant cover by fire, frost or agricultural practices leads to secondary re-establishment of vegetation, usually from residues of native seed and plant stock within the affected area. Secondary succession is generally a faster process than primary succession because not all organisms have been removed from the area following disturbance. Consequently, a secondary sere may complete its successional sequence in 50–300 years, whereas a primary sere may take 1000 years or more because

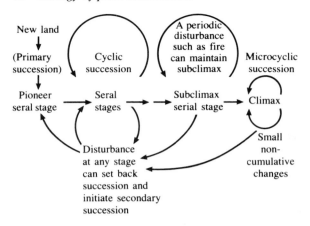

1.5 Diagrammatic pathway of different types of succession; primary, secondary and cyclic. The climax stage is in a state of dynamic equilibrium (after Barbour *et al.*, 1987).

the environmental conditions at the outset are more severe.

Differences in the rates of succession between one area and another may lead to a zonation of successional types. Some zonational categories have boundaries which change with time, whilst others are more permanent because one or a few environmental factors exert an overriding effect which prevents (or slows down) subsequent successional changes.

The major global plant formations (Ch. 4) may be considered as climatic climax communities. They represent vegetation types which have reached a stable community structure under the prevailing climatic conditions. Within these broad vegetational zones, however, local environmental factors (topography, geology and soils, drainage, human activity) may favour the development of other recognisable and apparently stable vegetation types.

Chapter 2

Soils

Soil is the weathered, uppermost layer of the Earth's land surface, which, when containing suitable amounts of air and water, can support vegetation. Soil depth varies from a few millimetres to several metres. Soil constituents can be divided into four groups: mineral particles and organic matter which make up the soil solids, and water and air which occupy the interstices between adjacent particles (Townsend, 1973).

2.1   Soil composition

2.1.1   *Mineral fraction*

All soils, apart from a few organic ones, contain a large proportion of mineral material that has been removed from the parent rock by denudation. This action takes place constantly by the processes of weathering and erosion (transportation). The changes brought about by weathering occur simultaneously and can be of a physical, chemical or biological nature.

Physical weathering results in the gradual decrease in the size of rocks without any of the mineral alterations associated with chemical weathering. The agents of physical weathering include water, ice, wind and temperature. Streams, rivers, glaciers and wind have a great abrasive action or 'cutting power', especially when laden with transported sediments, resulting in erosion as well as weathering. Frost action, the alternate freezing and thawing of water in crevices, exerts great pressure on the surrounding rock owing to the volume expansion of 9% that occurs when water freezes, resulting in the eventual cleavage and disintegration of rocks.

The most important agent of chemical weathering is rainwater containing dissolved carbon dioxide from the atmosphere. This dilute carbonic acid may bring about simple solution, hydration or, the most widespread and important process, hydrolysis. Simple solution removes calcium as soluble calcium bicarbonate from calcareous rocks, leaving behind insoluble material to be weathered by other processes. Hydration often accompanies hydrolysis and involves the combination of the ions H^+ and OH^- with a mineral to form a new compound, a process accompanied by swelling which exerts pressure on the parent rock. Hydrolysis is important in the weathering of many complex silicates, including feldspars and micas. An example is the hydrolysis of orthoclase feldspar $(K_2Al_2Si_6O_{26})$ to the clay mineral kaolinite $(Al_2O_3.2SiO_2.2H_2O)$ with potassium and surplus silica being washed away in solution.

Many plants have a biological role in the weathering of rocks. Lichens, for example, not only exert a weak solvent action on the rocks on which they live but they also maintain a thin layer of water at the rock surface that encourages the processes of chemical and physical weathering. Weathering produces rock particles of various sizes; these may remain at the site of weathering or be removed and deposited elsewhere.

2.1.2 *Organic matter*

The first stage in soil development involves the incorporation of organic material into the weathered mineral particles. This organic matter consists of fresh and partially decayed (i.e. humified) plant and animal remains as well as dark, amorphous material known as humus. Humus is an organic colloid with an organization similar to that of clay (section 2.2.3) but composed of carbon, hydrogen and oxygen rather than of silicon, aluminium and oxygen. Individual humus particles are very small and have a high cation exchange capacity (section 2.2.4).

The amount of organic matter present varies: some sandy soils contain less than 1%, whereas peaty fenland soils may have more than 80%. Disintegration of organic tissue is aided by soil organisms such as nematodes, earthworms, fungi and bacteria. The products of decomposition are principally water and carbon dioxide. However, the process is also important as a source of phosphorus, sulphur and nitrogen-containing compounds. The more resistant and stable end-products of this decomposition constitute the humus fraction of the soil (Brady, 1974).

There are three broad categories of humus: mull, mor and moder. Mull humus is associated with the brown earth soil type (section 2.4.1.1). The litter layer of undecomposed plant remains is thin and the decomposed organic material becomes rapidly incorporated into the surface layers of the soil. This decomposition is aided by a diverse soil flora and fauna, including earthworms which disperse the humus particles throughout the upper soil layers. Mull humus usually has a neutral or slightly alkaline pH. Mor or 'raw' humus has a much slower rate of decomposition of organic material, so that a deep litter layer may overlie the humified layer. This is a consequence of the strongly acidic nature of the humus, which both depresses the activity of decomposer bacteria and explains the almost total absence of earthworms. Mixing of organic and mineral matter is therefore reduced. The associated soil type is the podzol (section 2.4.1.2). A moder humus represents the midway stage in the transition between mull and mor humus. Superficially it resembles a mull, but it is acidic in reaction and there is no intimate mixing of organic and mineral matter.

2.1.3 *Soil atmosphere*

Soil air is of a different composition to that of the atmosphere. The nitrogen content is approximately the same, but there is usually more carbon dioxide and less oxygen making up the remainder. The soil carbon dioxide concentration may be 10–100 times greater than that of the atmosphere (c. 0.03%) but in dry soils is usually about 0.05%. Decomposition of organic material by micro-organisms leads to the presence of small amounts of the gases methane and hydrogen sulphide. When the oxygen level decreases in a soil, as occurs during waterlogging, then the gases methane, ethylene, ammonia and hydrogen sulphide, which can be injurious to plants even when present in trace amounts, increase.

2.1.4 *Soil water*

Water exerts a major influence on soil through leaching and evaporation, which lead to the development of soil horizons (section 2.3). Water in soils is bound to solid particles by matric forces, including adsorption on to

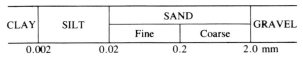

CLAY	SILT	SAND		GRAVEL
		Fine	Coarse	
0.002	0.02	0.2	2.0 mm	

2.1 Size classification of soil mineral particles according to the system of the International Society of Soil Science.

clay particles and capillary attraction in the soil pores or interstices.

2.2 Physical and chemical properties of soils

2.2.1 *Soil texture*

The texture of a soil is determined by the size of the soil particles (Fig. 2.1) and variations in the nature and amount of organic and inorganic colloidal material present within it. Soil texture influences drainage, aeration and moisture retention, and it also has an indirect effect on soil temperature through variation in soil moisture content – wet soils are colder and take longer to absorb heat from the sun than dry soils, which absorb heat faster. The proportion of air to water between the soil particles decreases with decrease in particle size, so that clays, which are fine-textured, have a high water content and low air content, which, in turn, can inhibit root respiration and water absorption. On the other hand, sandy soils, which have larger mineral particles, contain more air than clay soils; however, they are prone to desiccation.

2.2.2 *Bulk density*

Since soil is a combination of mineral, organic, air and water components in varying proportions, determinations of the density of soil solids may give a misleading impression of the true nature of the soil. Bulk density is the weight of unit volume of a dry soil and the pore space of the soil is included in its estimation. Porous, loose-textured soils (for example, silt and clay loams) have low bulk densities, whilst compact soils (for example, sandy soils) have high values. High organic matter content in soils produces low bulk density owing to the light weight of dried organic material. Peats, therefore, have the lowest bulk densities of all soil types.

2.2.3 *Mineralogy of soil particles*

Soil mineral particles are of two types: primary and secondary minerals. The chemical composition of the former has been little altered by the processes of weathering. The latter are formed by the destruction of less resistant minerals or by the reprecipitation from solution of the products of weathering.

The sand and silt fractions of soils are dominated by coarse rock fragments and primary minerals, especially silicates (quartz is the dominant silicate). Small amounts of secondary minerals, such as salts, hydroxides and oxides may also occur. The weathering of primary minerals results in the release of atoms or groups of atoms, some of which will be taken up by plants while others will be leached from the soil. However, the majority will regroup to form fine, secondary minerals called clay minerals.

The clay fraction of a soil consists predominantly of very fine, colloidal particles with diameters of between 0.1 and 0.001 μm. As a result of their small size, clay particles have a very large surface area per unit mass.

The crystalline organisation of clays varies from one type to another and this variation is reflected in their physical and chemical properties. Clays of temperate regions of the world have a layer-lattice structure, composed of alternating sheets of silica tetrahedra and aluminium octahedra with oxygen or hydrogen bonding between them. Owing to the surplus valencies of some oxygen atoms, the clay particles are negatively charged, and cations (in particular Al^{3+}, Fe^{3+}, Fe^{2+}, Mg^{2+}, Ca^{2+}, K^+, Na^+, NH_4^+ and H^+) and water molecules are adsorbed on to the clay crystals. As a result of their large surface area, clay particles have a high cation exchange capacity.

2.2.4 *Cation exchange capacity*

The adsorption of cations in solution by a soil colloid and the subsequent release of other cations held on the surface of that colloidal particle is termed cation exchange. The cation exchange capacity (CEC) of a soil is the total number of cations (milli-equivalents) that can be adsorbed per unit weight of soil. The CEC of a soil forms a reservoir of nutrients that are available to plants and micro-organisms. Soils, particularly those in regions of high rainfall, would rapidly become depleted of nutrients by leaching without this ability for cation adsorption on to colloidal surfaces. Cation exchange is exhibited both by mineral (clay) and organic (humus) colloids.

2.3 The soil profile

Five major factors influence soil formation: climate; the nature of the soil parent material; living organisms; topography; and the length of time that soil-forming (pedogenic) factors have been active. The interaction between these factors results in a variety of soils that

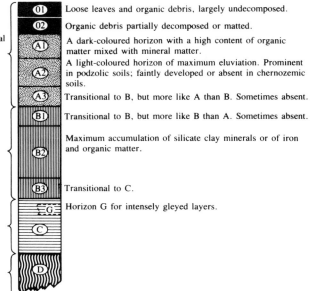

The weathered parent material (occasionally absent)

THE SOLUM
(The genetic soil developed by soil-forming processes.)

Horizons of maximum biological activity, of eluviation (removal of materials dissolved or suspended in water), or both.

Horizons of illuviation (accumulation of suspended material from A) or of maximum clay accumulation.

Any stratum underneath the soil, such as hard rock or layers of clay or sand, that is not parent material but which may have significance to the overlying soil.

O1   Loose leaves and organic debris, largely undecomposed.

O2   Organic debris partially decomposed or matted.

A1   A dark-coloured horizon with a high content of organic matter mixed with mineral matter.

A2   A light-coloured horizon of maximum eluviation. Prominent in podzolic soils; faintly developed or absent in chernozemic soils.

A3   Transitional to B, but more like A than B. Sometimes absent.

B1   Transitional to B, but more like B than A. Sometimes absent.

B2   Maximum accumulation of silicate clay minerals or of iron and organic matter.

B3   Transitional to C.

G   Horizon G for intensely gleyed layers.

2.2   Idealised soil profile (after Cruise & Newman, 1973).

may be identified by reference to their soil horizons. A soil horizon is a distinctive horizontal band that differs both in appearance and character from the adjacent bands. Horizons may be classified as O, A, B, C, or D. Other soil horizon classifications also exist and synonymous terms are indicated in the text and related figures. A diagrammatic representation of a soil profile appears in Fig. 2.2.

2.3.1   *O horizons*

These horizons form above the mineral soil and consist exclusively of organic matter. They are classified according to the amount of decomposition that has occurred. The O1 (A_{00}) horizon contains fresh litter (undecomposed plant remains), whereas the organic matter of the underlying O2 (A_0) horizon is partially or well decomposed. The development of the O2 horizon is affected by several factors, in particular the climate. In conditions of low temperature and anaerobism the microbial activity within the soil is depressed, so that an accumulation of acid mor humus is favoured at the surface. In drier and warmer conditions a shallower mull humus horizon develops.

2.3.2   *A horizons*

These mineral horizons lie at or near the soil surface and are enriched with organic matter. They are characteristically influenced by eluviation (leaching), which involves the downwards movement of water that carries the soluble decomposition products from the organic horizons above. These impart an acid pH to the water, increasing its solvent action, so that soluble salts, clays, and iron and aluminium oxides are removed to lower horizons. Subdivisions of the A horizon are as follows:

A1   horizon: contains dark, humified organic matter incorporated with the mineral material.

A2   (E_a) horizon: zone of maximum eluviation in which there may be a concentration of resistant minerals such as quartz. This horizon may be light in colour, as if bleached, a feature that is particularly prominent in podzolic soils (section 2.4.1.2).

A3   horizon: not always present; transitional between the A and B horizons.

2.3.3   *B horizons*

These represent the zone of illuviation or accumulation of leached constituents such as clays, iron and aluminium oxides and organic material. The B layer often has

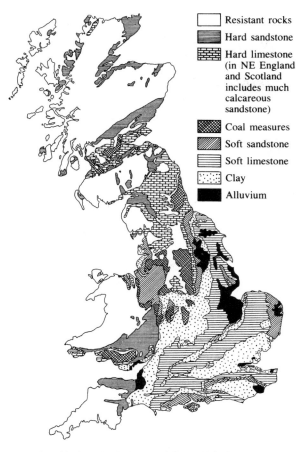

Resistant rocks

Hard sandstone

Hard limestone (in NE England and Scotland includes much calcareous sandstone)

Coal measures

Soft sandstone

Soft limestone

Clay

Alluvium

2.3 Simplified geological map of Great Britain.

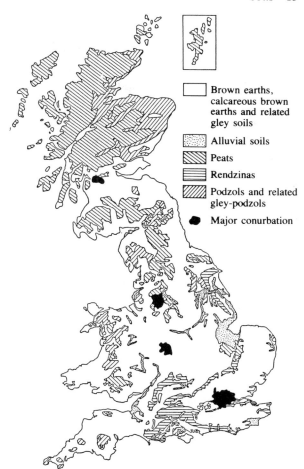

Brown earths, calcareous brown earths and related gley soils

Alluvial soils

Peats

Rendzinas

Podzols and related gley-podzols

Major conurbation

2.4 Simplified soil map of Great Britain.

a higher bulk density than other horizons as a result of this accumulation. Furthermore, mineral particles may become concreted together into clumps and layers known as pans. These are particularly common in podzols (section 2.4.1.2) where they may prevent free drainage. Three subdivisions may be recognised:

B1 horizon: similar to the A3 horizon, but has more characteristics of B rather than A horizons. It may be enriched with iron oxides and clay.

B2 horizon: a lower zone enriched with clays (indicated by the notation B_t), sesquioxides (B_s), iron (B_i) and humus (B_h).

B3 (B/C) horizon: transitional between the B and C horizons.

2.3.4. *C and D horizons*

The C horizon is the underlying mineral material from which the A and B horizons have developed. The D horizon, or bedrock, underlies the C horizon and is not necessarily the source of the overlying A, B and C horizons. For example, glacial drift may overlie the true bedrock.

2.4 Soil types of the British Isles

In Britain and western Europe the two major zonal (climatic) soils are brown earths and podzols. These are as much a function of the climate, in particular rainfall, humidity and temperature, as of the vegetation that they support. Besides these two zonal soils, there are several other soil types which reflect a local, dominant pedogenic factor. These soils are intrazonal and include gley and peat soils of waterlogged areas and rendzinas

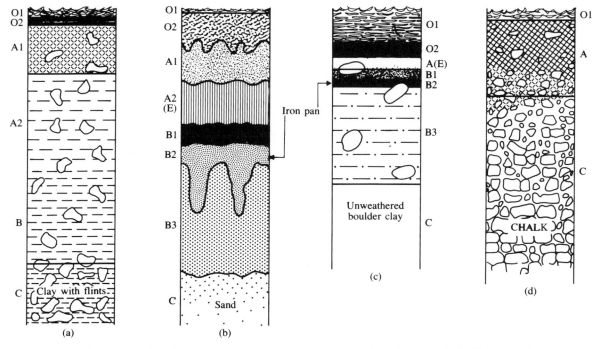

2.5 Brown earth, podzol and rendzina profiles. (a) Brown earth soil in a Chiltern beechwood, the B (illuvial) layer is poorly defined and there is a gradation from the A2 to the C horizon. (b) Podzol over sand with a thick surface mor horizon forming compacted peat below, bleached A2 elluvial horizon and a prominent undulating iron pan. (c) Podzol overlying boulder clay – there is a thick O1 horizon of poorly decomposed litter, underlain with peat. The elluvial (A) and illuvial (B) horizons are thin and there is a well-developed iron pan. (d) Rendzina over chalk. The blackened A horizon grades directly into the limestone C horizon. (Modified from Tansley, 1939.)

formed over calcareous bedrock. Azonal soils are strongly influenced by erosional or depositional processes and are without horizon differentiation; most soils developed on recent alluvial deposits are of this type, as are regosols and lithosols. Azonal soils lack the characteristics of the zonal soils for that particular climatic region. (Curtis *et al.*, 1976).

2.4.1 *Zonal soils*

2.4.1.1 *Brown earths*

Brown earths (Fig. 2.5) are characteristic of the temperate deciduous forest zone under conditions of low or moderate rainfall (<1000 mm) (hence their synonym 'brown forest soils'). They typically form on siliceous parent materials such as shales, fine sandstones, clays and also impure limestones. Litter is broken down rapidly at the soil surface so that the O2 horizon is either shallow or non-existent. The underlying, well-aerated A

horizon consists of nutrient-rich mull humus, in which the large variety of soil micro-organisms and animals mix the soil material from different horizons. Consequently, the B horizon is rarely delimited clearly. The brown earth profile is represented as A(B)C and has a slightly acid pH (5–6.5).

Under conditions of higher rainfall carbonates and clays may be removed from the upper horizons so that the pH drops further into the acid range. Although these leached brown earths may start to resemble podzols owing to the presence of a definite B horizon, the humus is still of the mull or moder type.

2.4.1.2 *Podzols*

Podzols–from the Russian *pod* = under and *zol* = ash, referring to the colour of the A2 horizon–are usually found overlying the hard rocks of the north and west of the British Isles, where they are formed under the influence of intense eluviation and illuviation. In a European context podzols are characteristic of the

Boreal Forest belt (taiga—section 4.2) where the climate is cool and wet and the acid mor humus (pH 4–5) supports coniferous forest, or heather moorland if the trees have been removed. Occasionally, podzols also develop further south in drier, warmer climes, but only on nutrient-poor, coarse sands and gravels which are easily leached.

A well-developed podzol profile has both O1 and O2 horizons overlying the A1 mineral horizon. Although the A1 horizon is usually stained dark-grey or black by organic matter from the surface mor layer, the individual silt and sand particles within it are light-coloured, as if bleached. This results from the removal of their mineral coatings, in particular ferric sesquioxides, by the acidic, percolating water. The A2 horizon shows the same type of eluviation, but without the organic staining, so that it is typically grey or even white in colour. Powerful eluviation causes the A horizons to become depleted of ferric, manganic and aluminium sesquioxides, clay minerals, Ca^{2+}, Mg^{2+} and K^+. As a result they are nutrient-deficient.

Beneath the bleached A horizons there is a sharp transition to the illuvial B horizons where the accumulation of eluviated substances occurs. The upper B1 horizon is dark-brown and enriched with organic matter and ferric and aluminium sesquioxides, whereas the lower B2 horizon is reddish brown and enriched only with ferric sesquioxide. These products of eluviation are mainly deposited as coatings on mineral particles which have a tendency to concrete together to form 'hard pans' or 'iron pans'. If impervious to water these indurated layers can cause waterlogging of the overlying soil, conditions become reducing, organic matter decomposition slows down and peat may accumulate (section 2.4.2.3). The podzol profile is OABC in composition.

2.4.2 *Intrazonal soils*

2.4.2.1 *Rendzinas*
The thin soils that develop on relatively pure limestones, especially chalk, are termed rendzinas (from the Polish for a rich, limey soil) (Fig. 2.5). These are rather immature soils with a shallow A horizon (up to 0.35 m in depth) resting directly on the calcareous bedrock of the C horizon. The black humus is of the mull type and, owing to the mixing of weathered limestone particles by earthworms, has a high calcium carbonate content and a pH of 7 to 8. Consequently, acid by-products of

decomposition are neutralised rapidly so that there is little or no leaching and percolating water drains away quickly through the permeable bedrock. The full profile is therefore only AC.

Impure limestones contain a large proportion of insoluble sands, silts and clays which accumulate as the surrounding calcium carbonate is removed in solution. Deep, rich soils develop in these areas, especially where the land is fairly flat. Since the A horizon of these soils has a low calcium carbonate content, the percolating water is acidic and leaching is encouraged. The colour of the A horizon is lighter than that of the rendzina and the mull humus layer is slightly acid or neutral. The B horizon is reddish brown and lies directly on the C horizon. The resultant soil is called 'calcareous brown earth', the profile of which is ABC.

2.4.2.2 *Gley soils*
Gley (from the Russian for clay) soils (Fig. 2.6) have some of their horizons waterlogged for part or all of the year and these are greatly affected by the associated anaerobic conditions, which result in the chemical reduction of iron and other elements. Where there is a soil oxygen deficiency iron is in the ferrous state and imparts a blue or grey colour to the soil. When, however, the ground water rises high enough to meet air-filled cracks or living roots then the ferrous iron is oxidised and ferric hydroxide is precipitated as a yellow, red or rust-brown deposit in the profile. This gley horizon represents a zone of alternating oxidising and reducing conditions which usually appears as a band of mottled colours in the soil. When describing a soil profile in which gleying is apparent the suffix 'g' is placed after the horizon symbol, e.g. B_g. Associated with low soil oxygen is the retarded decomposition rate of organic matter, which either leads to organic enrichment of the A horizon or, where waterlogging is more or less constant, the build-up of a surface peat layer.

There are two situations in which a gley soil can occur: surface water gleys develop at sites where water is retained at the surface by an impervious layer in the soil, whereas ground water gleys are characteristic of sites with high ground water levels. Gley soils are well represented throughout the British Isles.

2.4.2.3 *Peats*
As indicated in sections 2.4.1.2 and 2.4.2.2 waterlogging of the surface layers of the soil horizon may lead to the accumulation of organic material which, if continued

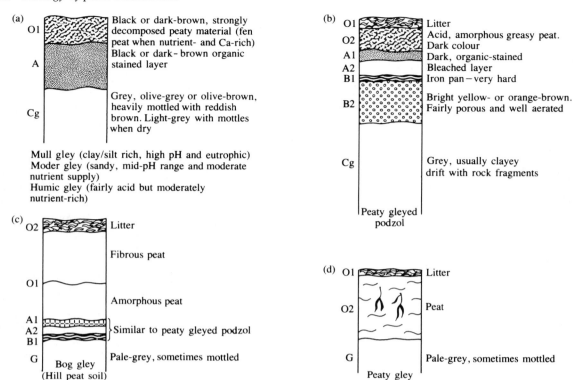

2.6   Formation of gley soils under different conditions of soil wetness, acidity and nutrient status: (a) mull gley; (b) peaty gleyed podzol; (c) bog gley; (d) peaty gley (from Etherington, 1975).

over a long period of time, results in the development of a peat horizon up to several metres thick (section 7.2.3). Peat can also form in lake basins as a result of hydroseral infilling (sections 6.4 and 7.1). The nature of peat varies according to the dominant plant species present at the different stages of the hydroseral succession. Peat can be referred to as sedimentary, fibrous or woody depending on whether or not the parent plants were those of open water (e.g. pondweeds, water lilies or plankton), submerged and emergent species of transitional successional phases (e.g. sedges, reeds and bryophytes, especially *Sphagnum* spp.) or trees and shrubs of the climatic climax forest (e.g. species of *Pinus, Salix, Frangula, Betula* and *Alnus*).

Moore and Bellamy (1973) distinguish three different types of peat according to the water supply available at various stages of the peat-forming process (Fig. 2.7). In order to distinguish these distinct edaphic conditions they propose the terms primary, secondary and tertiary peat as follows:

1. Primary peats are formed in waterlogged basins or depressions and the peat accumulates in the water body.
2. Secondary peats develop above the original water table limit of the basin or depression and in so doing they elevate the water table above the original level. Although the surface of the peat may dry out at periods of low water supply, ground water flooding

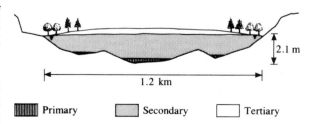

2.7   Section across a raised mire showing the relative positions of primary, secondary and tertiary peat (from Moore & Bellamy, 1973).

2.8 (a) View of quartzite scree on Beinn Eighe, north-west Scotland. These unstable accumulations of rock are an example of a lithosol. (b) Regosol of unconsolidated dune sand at Newborough Warren, Anglesey.

or water table elevation occurs with sufficient regularity to maintain the minerotrophic nature of the peat and its surface vegetation. (1) and (2) are the rheotrophic mires referred to in Chapter 6.

3. Tertiary peats develop above the physical limits of the ground water by the ability of *Sphagnum* moss to continue its upward growth utilizing rainfall as its sole water and nutrient source. These are the ombrotrophic mires referred to in Chapter 7.

2.4.3 *Azonal soils*

2.4.3.1 *Alluvial soils*

The mineral alluvium which covers many flat valley bottoms gives rise to alluvial soils which lack a well-differentiated profile. Marsh soils do not have obvious horizons and are derived from clay or silt alluvium of either freshwater or marine origin. Riverside meadow soils are maintained in an immature condition as a result of the frequent input of silt during flooding and a constantly fluctuating water table. If the flooding of these soils has not been regulated by man then their fertility and humus content are usually very high.

2.4.3.2 *Regosols and Lithosols*

Regosols (blanket soils) in which horizons do not develop occur on unconsolidated, soft, mineral deposits of a fine texture such as glacial till, loess and dune sand (Fig. 2.8). Lithosols (stone soils) are derived from harder substrates so that the soil consists of coarse, partially weathered rock fragments. Tundra stony soils and unstable screes are of this type. The profile of these 'raw soils' may be classed as (A)C. Soils with a slightly better developed profile may also be classified as lithosols, for example, some thin, stony rendzinas and rankers which have an AC profile. The ranker profile consists, at best, of a virtually unweathered C horizon derived from nutrient-deficient, siliceous bedrock and a thin, superficial A horizon which contains very little humus. Rankers are characteristic of cold tundra, high mountain areas and semi-arid regions.

Chapter 3

Phytosociology

Phytosociology (also known as plant sociology or phyto-cenology) is the science of vegetation and involves investigation of all characteristics of plant communities: physiognomy, floristic composition, community morphology, structure, development and change, multilateral relations of plants to one another and to the environment, and the classification of communities (Becking, 1957). Phytosociology is based on two features of plant communities: (1) plants are not randomly distributed, rather there are distinct species combinations which repeat themselves regularly in nature; (2) the complex interaction between plants and habitat (environment) and between individual plants.

There are several schools of phytosociology which place different degrees of emphasis on floristic composition, species dominance and environmental factors. In Europe the most widely used methods of vegetation description and classification are those of the Zurich-Montpellier school (or Braun-Blanquet school) (Braun-Blanquet, 1928). This approach is being used increasingly in Britain. It is founded on the development of a system of classification based upon an inventory of vegetation types. The fundamental unit of this vegetation classification is the association, which is an abstract generalisation derived following comparison of a large number of species lists made at selected field sites.

Phytosociology is an invaluable method for vegetation survey and assessment. It provides a framework within which plant community structure and succession can be explained, whilst the field techniques are simple and rapid to employ. The phytosociological approach of the Zurich-Montpellier school has been applied to all the vegetation types described in this book.

3.1   Methodology

Associations are defined entirely by floristic composition, *not* by habitat; but they are also regarded as ecological realities (i.e. occupying definite habitats) (Poore, 1955; Shimwell, 1971).

Shimwell (1971) identifies five steps involved in the elucidation and characterization of associations:

1. field description;
2. aggregation of field data into tables to represent local variations in vegetation;
3. checking ecological reality of units extracted in the field;
4. investigation of similar patterns in other localities (to obtain the overall pattern of variation within a particular vegetation type); and
5. the erection, differentiation and characterization of associations.

Successively higher units are the alliance, order and class (see Table 3.3). The most complex is the vegetation circle, which is synonymous with plant formation (Ch. 4).

The field technique involves the subjective selection of sample stands of vegetation for description (French *relevés*, German *Aufnahmen*), which should be 'uniform', i.e. floristically homogeneous. These terms are used to designate separate plant aggregations occupying a certain space, belonging to a stated association and homogenous to an extent that makes it difficult to divide them into smaller parts differing from one another, either in their dominant species, or in the different arrangement of their components. The area of a single stand of an association can vary greatly: from several square centimetres to several, or even several hundred thousand, square metres.

The area of the sample plot is variable and depends on the plant community being analysed. Becking (1957) lists two requirements for the size of the sample plot area: (1) that it permits inclusion of only the most typical segments of homogeneous vegetation; (2) that the size be at least equal to, but preferably greater than, that of the minimal area of the community under study. Although the minimal area can be calculated mathematically, ecological experience is usually more important than a statistically proven minimal area. In the absence of any statistically based sampling method, the overall appearance of the vegetation is likely to dictate to the observer which stands should be selected.

Following sample plot selection, site description data are collected. These comprise basic environmental and geographical information (e.g. locality, grid reference, aspect, altitude, slope, soil type, etc.) and a list of all plants present with a numerical estimation of their cover-abundance and sociability, both of which are expressed using five-point scales. The first scale relates the abundance (i.e. density) of individual species in the relevé to their cover (i.e. the percentage of the total plot area covered by the aerial vegetative parts of that species). Sociability is related to the spatial distribution of individuals of the same species and is dependent not only on the growth form of a plant, but in a much greater measure on environmental conditions (Table 3.1).

To a large extent sociability is the expression of the degree of success of a plant in a given environment. Cover- abundance and sociability indices are placed after a species name when recording a relevé (the cover-abundance digit being given first).

The next stage in this method involves the tabulation of data from similar vegetation stands and the synthesis of vegetation units. Vegetation tables are composed of a two-dimensional array of relevés and species: relevés are

Table 3.1 Phytosociology measures: (a) Braun-Blanquet scale of cover/abundance; (b) Braun-Blanquet scale of sociability; (c) percentage constancy classes.

(a) Cover-abundance	(b) Sociability	(c) Percentage constancy classes
+ Occasional	1 Growing singly	r < 1% presence in relevé
1 <5%	2 Growing in small groups of a	I 1–20%
2 5–25%	few individuals	II 21–40%
3 25–50%	3 Large groups of many	III 41–60%
4 50–75%	individuals	IV 61–80%
5 75–100%	4 Patches or a broken mat	V 81–100%

	5 Extensive mat completely
	   covering (or almost) the whole
	   plot area	

Table 3.2 Phytosociology table for selected *Cratoneuron commutatum* springs in the southern Pennines.

Reordered number	1	2	3	4	5	6	7	8	9	10	11		
Relevé Reference No.	c2	c6	c19	c7	c1a	c10	c20	c24	c22	c23	c4		
Altitude (m)	380	380	385	380	380	365	380	335	335	330	245		
Slope (°)	80	70	80	95	80	75	70	85	80	75	80		
Flow rate	S	S	S	S	S	S	S	S	S	S	S	= slow	
Shade (%)	0	0	0	0	0	0	0	0	0	0	30		
Plant cover (%)	95	90	85	95	95	90	100	90	95	90	95		
Water pH	7.5	7.6	7.7	7.6	7.5	8.0	7.7	7.3	7.2	7.0	7.4		
Substrate	S	S	S	S	S	S	S	S	S	S	S	= silt	
Aspect	S	N	NE	N	S	NE	N	N	N	S	S		
Quadrat size (m²)	1	1	1	1	1	1	1	1	1	1	1		

Species	1	2	3	4	5	6	7	8	9	10	11	CONSTANCY	
Cratoneuron commutatum	2–4	3–4	3–3	3–4	4–4	3–4	3–3	3–4	3–3	2–3	3–4	V$^{2-4}$	100%
Bryum pseudotriquetrum	2–3	2–4	+	1–3		2–3	1–3		1–3	1–3	1–3	V^{+-2}	82%
Calliergon cuspidatum						1–3	1–3	1–3	1–2	1–3		III^{1}	46%
Philonotis fontana	1–2							1–3	1–3	1–3		II^{1}	36%
Cardamine pratensis								+	+	1–1		II^{+-1}	27%
Sagina procumbens								1–1		1–2		I^{1}	18%
Epilobium palustre								+	1–2			I^{+-1}	18%
Equisetum palustre						1–4				1–2		I^{1}	18%
Angelica sylvestris										1–1		r	
Geranium robertianum										+		r	
Heracleum sphondylium										1–1		r	
Tussilago farfara										1–2		r	
Valeriana officinalis										+		r	
Festuca ovina	1–4	2–3		1–1	1–4	1–4		1–4	1–4	1–4		IV$^{1-2}$	73%
Pellia epiphylla	2–3	+	2–4				2–3	1–2		1–2	1–2	IV^{+-2}	64%
Chrysosplenium oppositifolium			2–4		2–3	2–4	2–3	1–4		+		III^{+-2}	55%
Cirsium palustre		+		1–1			+		+			II^{+-1}	36%
Poa annua	+			1–2	1–1		+					II^{+-1}	36%
Agrostis stolonifera			1–4				1–4			+		II^{+-1}	27%
Holcus lanatus								1–1	+	+		II^{+-1}	27%
Crepis paludosa						1–1				3–4		I$^{1-3}$	18%
Ranunculus repens		+		+								I^{+}	18%
Brachythecium rivulare									1–4	1–2		I^{1}	18%
Marchantia polymorpha							+			+		I^{+}	18%

listed in columns and species in rows. Data tables can be rearranged to highlight the existence of mutually exclusive groups of species, i.e. groups of differential species, whose occurrence is restricted to certain groups of relevés. Re-ordering of the primary data table (or raw table) on the basis of these differential species produces a partially ordered table, in which relevés with similar floristic composition are placed close together. The final extract or association table (Table 3.2) displays groups or stands whose defined by the presence of particular character species, whose occurrence is restricted to the community represented by those stands. Association character species can only be assigned after the whole range of the community type under investi-

gation has been surveyed, and positioning of the association in the hierarchical floristic system of Braun-Blanquet (i.e. association, alliance, order, class) requires consideration of the sociological rank and ecological amplitude of each species (Becking, 1957; Moore, 1962; Shimwell, 1971).

Associations may also be divided into subordinate units. These lower units are usually characterized by differential taxa and include the sub-association, which might be based on local edaphic or micrometeorological differences. The next lowest rank is the variant, and below this the sub-variant may be used if required. The lowest unit is the facies, which is usually characterised by the dominance of one of the species belonging to the normal floristic assemblage of an association (Whittaker, 1973).

The criterion for the homogeneity of any plant community is the presence (constancy) of certain species. If a definite number of relevés are compared in a synthesis table, constancy of the species present can be calculated as a percentage of the total number of relevés (Table 3.1). The degree of constancy indicates the probability of encountering at least a single specimen of a certain species in any stand of the association selected at random. From an ecological point of view it denotes the percentage of stands of an association in which the given species is involved as a competitor. Frequently associations which are related in terms of floristics, physiognomy, structure and ecology are grouped together into a constancy table and this can provide information on the phytogeography of a region.

Objections raised against the Braun-Blanquet method concern both the field description procedure and the subsequent data analysis, and it has often been criticised for the subjectivity of the sampling procedure. There are also instances where the Braun-Blanquet method cannot be easily applied. For example, in vegetation where species are not conspicuously different in foliage from each other (e.g. in grasslands), making estimations of cover-abundance and sociability difficult. Likewise, problems arise if the minimal area is too large for visual estimates of cover to be made (e.g. owing to the heterogeneous nature of tropical rain forest vegetation, the minimal area may exceed 10 000 m²). The disadvantages of this procedure, however, have to be weighed against the advantages. The Braun-Blanquet method is simple, comparatively easy to apply and rapid in execution. The methods of the Zurich-Montpellier school offer an effective system of classification, which is the most widely applied and most international of all the phytosociological methods in the study of vegetation. In particular, they provide a standard technique whereby descriptions can be made in any kind of vegetation, and the results obtained by ecologists in different regions can be directly compared. (Moore *et al.*, 1970).

3.2   Nomenclature

For designating the separate associations use is made of international Latin names, formed by adding the suffix '-etum' to the radical of the Latin generic name of the plant that is chosen as the distinguishing one, e.g. Cardaminetum, Saxifragetum. To this is added the specific name in the genitive case (e.g. Cardaminetum amarae, Saxifragetum aizoidis). It is usual to choose as distinguishing plants, from which the names of associations are formed, the dominant species in the given association and/or its characteristic species. Higher units in the classification are delimited by adding the suffixes shown in Table 3.3 and a synopsis of the phytosociological classification of the class Montio-Cardaminetea is shown in Table 3.4.

Table 3.3 Nomenclature of phytosociological classificatory units.

Rank	Suffix	Example
Class	-etea	Montio-Cardaminetea
Order	-etalia	Montio-Cardaminetalia
Alliance	-ion	Cardamino-Montion
Association	-etum	Philonotido fontanae-Montietum
Sub-association	-etosum	Ranunculetosum aquatilis
Variant	—	(specific names used)

Table 3.4 Synopsis of the phytosociological classification of the Montio-Cardaminetea (derived from various sources).

Class	Order	Alliance	Sub-alliance	Association (selection only)
Montio-Cardaminetea	Montio-Cardaminetalia	Cardamino-Montion	Montion	Philonotido fontanae-Montietum
			Cardaminion	Cardaminetum amarae
		Mniobryo-Epilobion hornemannii		Pohlietum glacialis
	Cardamino-Cratoneuretalia	Cratoneurion commutati		Cratoneuretum commutati
		Cratoneureto-Saxifragion aizoidis		Saxifragetum aizoidis

3.3 Synopsis of the higher classificatory units of the British vegetation

The phytosociological units presented in this synopsis are based on the system devised by Braun-Blanquet (1928) as modified by various authorities, in particular Lohmeyer *et al*. (1962), Szafer (1966) and Westhoff and Den Held (1969). The classes have been arranged in ten groups which embrace the major vegetation types of the British Isles.

I. *Freshwater communities of free-floating, bottom rooted submerged, and bottom rooted floating-leaved plants*

Class 1 Charetea (Chapter 6)
Submerged deep-water communities (4–5 m) dominated by algae of the Charophyta (stoneworts)
  O. Charetalia
    A. Charion fragilis
    A. Charion canescentis

Class 2 Lemnetea (Chapter 6)
Free-floating duckweeds of mesotrophic and eutrophic waters
  O. Lemnetalia
    A. Lemnion minoris

Class 3 Potametea (Chapter 6)
Bottom rooted angiosperm associations of mesotrophic, eutropic and brackish waters
  O. Magnopotametalia — tall, submerged pondweeds and floating-leaved plants
    A. Magnopotamion
    A. Nymphaeion
  O. Parvopotametalia — small, mainly submerged plants
    A. Parvopotamion
    A. Hydrocharition
    A. Callitricho-Batrachion
  O. Luronio-Potametalia — submerged associations of shallow acidic waters
    A. Potamion graminei

Class 4 Littorelletea (Chapter 6)
Rooted aquatic vegetation of the littoral zone of oligotrophic and dystrophic lakes
  O. Littorelletalia
    A. Littorellion uniflorae

II. *Emergent vegetation of freshwater swamps, springs and flushes*

Class 5 Phragmitetea (Chapter 6)
Reed-grass and tall sedge vegetation of shallow water in lakes, rivers, canals and coastal fresh and brackish water marshes
  O. Nasturtio-Glycerietalia — eutrophic drainage channels and shallow pools
    A. Glycerio-Sparganion
    A. Apion nodiflori
  O. Phragmitetalia — tall reed-swamps
    A. Phragmition
    A. Oenanthion aquaticae
  O. Magnocaricetalia — tall grass and sedge beds
    A. Magnocaricion

Class 6 Montio-Cardaminetea (Chapters 6 and 9)
Spring-heads and flushes of varying trophic status
  O. Montio-Cardaminetalia — oligotrophic springs
    A. Cardamino-Montion
    A. Mniobryo-Epilobion
  O. Cardamino-Cratoneuretalia — calcareous and minerotrophic springs
    A. Cratoneurion commutati
    A. Cratoneureto-Saxifragion aizoidis

Class 7 Parvocaricetea (Chapters 6, 8, and 9)
Low-growing sedge communities of transition mires, calcareous fens and minerotrophic flushes
  O. Caricetalia nigrae — mires and mesotrophic flushes
    A. Caricion curto-nigrae
  O. Tofieldietalia — calcareous fens and minerotrophic flushes
    A. Eriophorion latifolii

III. *River-bank willow woodland, fen and bog carr*

Class 8 Alnetea glutinosae (Chapter 6)
Alder swamps on organic substrates

O.   Alnetalia glutinosae
   A.   Alnion glutinosae

Class 9   Franguletea (Chapter 6)
Alder buckthorn and willow carr
  O.   Salicetalia auritae
     A.   Salicion cinereae

Class 10   Salicetea purpureae (Chapters 6 and 8)
River bank willow scrub
O.   Salicetalia purpureae
     A.   Salicion albae

IV.   *Acid bog and wet heath*

Class 11   Scheuchzerietea (Chapter 7)
Sphagnum-dominated communities of peat bog pools and hollows
  O.   Scheuchzerietalia palustris
     A.   Rhynchosporion albae

Class 12   Oxycocco-Sphagnetea (Chapter 7)
Peat-forming vegetation of ombrotrophic mires and wet heaths
  O.   Ericetalia tetralicis — wet heath
     A. Ericion tetralicis
  O.   Sphagnetalia magellanici — hummock-forming Sphagnum communities
     A.   Erico-Sphagnion
     A.   Sphagnion fusci

V.   *Tall shrub, scrub and climax forest*

Class 13   Rhamno-Prunetea (Chapter 8)
Scrub of woodland edge and stabilized sand dunes
  O.   Prunetalia spinosae
     A.   Berberidion
     A.   Salicion arenariae

Class 14   Vaccinio-Piceetea (Chapter 5)
Pine and birch woodlands
  O.   Vaccinio-Piceetalia
     A.   Vaccinio-Piceion
     A.   Betulion pubescentis

Class 15   Quercetea-robori-petraeae (Chapter 5)
Acid oakwoods
  O.   Quercetalia-robori-petraeae
     A.   Quercion-robori-petraeae

Class 16   Querco-Fagetea (Chapter 5)
Lowland mixed woodlands on nutrient-rich soils
  O.   Fagetalia sylvaticae
     A.   Alno-padion
     A.   Carpinion betuli
     A.   Fagion sylvaticae
     A.   Ulmion carpinifoliae

Class 17   Betulo-Adenostyletea (Chapter 9)
Montane tall herb and shrub communities
  O.   Adenostyletalia
     A.   Lactucion alpinae

VI.   *Grasslands and dry heaths*

Class 18   Nardo-Callunetea (Chapters 5, 8 and 9)
Species-poor acid grasslands and heaths
  O.   Nardetalia — grasslands
     A.   Eu-Nardion
     A.   Nardo-Galion saxatilis
     A.   Nardeto-Caricion bigelowii
  O.   Calluno-Ulicetalia — heaths
     A.   Calluno-Genistion
     A.   Empetrion nigri

Class 19   Molinio-Arrhenatheretea (Chapter 5)
Hay meadows, permanent pastures and adjacent footpath and field margins
  O.   Molinietalia — damp grasslands
     A.   Calthion palustris
     A.   Filipendulion
     A.   Junco-Molinion
  O.   Arrhenatheretalia — dry grasslands
     A.   Arrhenatherion elatioris
     A.   Cynosurion cristati

Class 20   Festuco-Brometea (Chapter 5)
Southern chalk and limestone grasslands
  O.   Brometalia erecti
     A.   Xerobromion
     A.   Mesobromion

Class 21 Elyno-Seslerietea (Chapter 9)
Northern montane grasslands and grass heaths overlying calcareous substrata
O. Elyno-Dryadetalia

Class 22 Loiseleurio-Vaccinietea (Chapter 9)
Tundra grasslands and grass heaths of high altitude and latitude
O. Caricetalia curvulae
    A. Loiseleurieto-Arctostaphylion

Class 23 Salicetea herbaceae (Chapter 9)
Chionophilous dwarf willow and bryophyte communities
O. Salicetalia herbaceae
    A. Cassiopeto-Salicion herbaceae
    A. Ranunculeto-Anthoxanthion

VII. *Brackish water and salt marshes*

Class 24 Zosteretea (Chapter 8)
Eel-grass communities of lower tidal mud flats
O. Zosteretalia
    A. Zosterion marinae

Class 25 Spartinetea (Chapter 8)
Cord-grass salt marshes
O. Spartinetalia
    A. Spartinion

Class 26 Thero-Salicornietea (Chapter 8)
Glasswort meadows of coastal mud flats
O. Thero-Salicornietalia
    A. Thero-Salicornion

Class 27 Ruppietea (Chapter 8)
Coastal brackish ditches and ponds
O. Ruppietalia
    A. Ruppion maritimae

Class 28 Asteretea tripolii (Chapter 8)
Species-rich grass and herb vegetation of the supra-littoral zone
O. Glauco-Puccinellietalia
    A. Puccinellion maritimae
    A. Armerion maritimae
    A. Puccinellio-Spergularion salinae
    A. Halo-Scirpion

VIII. *Coastal strand, shingle beach and sand dune*

Class 29 Cakiletea maritimae (Chapter 8)
Coastal strand and maritime edge communities
O. Thero-Suaedetalia
    A. Thero-Suaedion
O. Cakiletalia maritimae
    A. Atroplicion littoralis
    A. Salsolo-Honkenyion peploidis

Class 30 Ammophiletea (Chapter 8)
Pioneer vegetation of sand hills dominated by rhizomatous grasses and ephemeral herbs
O. Elymetalia arenarii
    A. Agropyro-Honkenyion peploidis
    A. Ammophilion borealis

IX. *Rocks, walls and scree*

Class 31 Asplenietea rupestris (Chapter 10)
Fern-dominated communities of rocks and walls
O. Tortulo-Cymbalarietalia
    A. Parietarion judaicae
    A. Cymbalario-Asplenion

Class 32 Thlaspietea rotundifoliae (Chapter 9)
Alpine and sub-alpine scree
O. Androsacetalia alpinae
    A. Androsacion alpinae

X. *Ruderal weed vegetation of waste places, spoil tips, arable fields and gardens*

Class 33 Plantaginetea majoris (Chapter 10)
Pioneer communities of open, disturbed, anthropogenic substrates
O. Plantaginetalia majoris
    A. Lolio-Plantaginion
    A. Agropyro-Rumicion crispi

Class 34 Artemisietea vulgaris (Chapter 10)
Nitrophilous tall herb communities of marginal areas and relatively stable, formerly disturbed ground

O. Artemisietalia vulgaris
   A.   Arction
   A.   Galio-Alliarion
   A.   Aegopodion podagrariae

Class 35 Epilobietea angustifolii (Chapter 10)
Nitrophilous communities of woodland edge and disturbed urban sites on a variety of soils
O. Epilobietalia angustifolii
   A.   Epilobion angustifolii

Class 36 Chenopodietea (Chapter 10)
Nitrophilous weed communities of rootcrop fields and pioneer vegetation in the built environment
O. Polygono-Chenopodietalia
   A.   Polygono-Chenopodion
O. Sisymbrietalia
   A.   Sisymbrion
   A.   Polygono-Coronopion

Chapter 4

The major plant formations of the world

Geographically widespread plant communities with similar physiognomy and life-form make up the largest vegetational units of the biosphere – plant formations, the boundaries of which are determined, primarily, by major world climatic differences. Temperature and rainfall are the most important climatic factors limiting the distribution of the indigenous species within each plant geographical region (Fig. 4.1). To a lesser degree, differences in soil type and surface topography are also

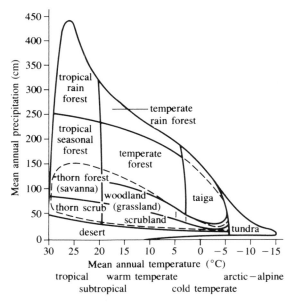

4.1 A pattern of world plant formation types in relation to rainfall and temperature. Boundaries between types are approximate. The dot and dash line encloses a wide range of environments in which either grassland, or one of the types dominated by woody plants, may form the prevailing vegetation in different areas (redrawn from Whittaker, 1970).

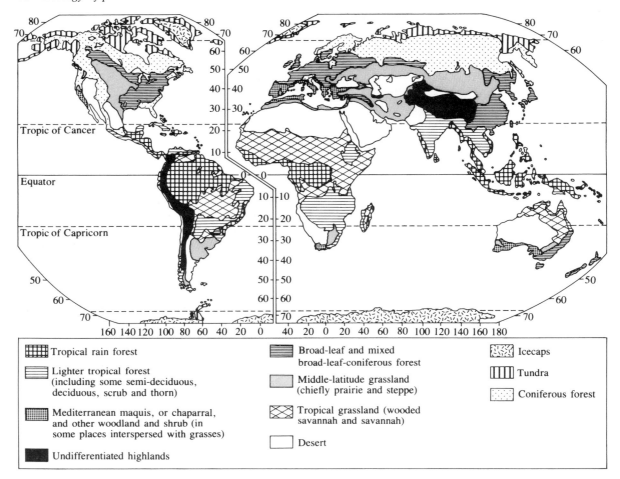

4.2   Distribution of the major plant formations of the world (after Walter, 1973).

important. Although the plant formations described in this chapter are examples of climax vegetation delimited by essentially natural phenomena, many have been modified, at least in part, by man's activities and only in the most extreme environments can the anthropogenic factor be virtually eliminated (e.g. in polar, desert or high mountain situations). Other plant formations have suffered from human exploitation, e.g. the use of forests for timber production and the intensive grazing and cultivation of grasslands. The distribution of the major plant formations of the world is illustrated in Fig. 4.2.

Certain ecosystems are more fragile than others. The conversion of temperate deciduous forest to productive arable land or pasture has been, for the most part, a successful process. However, in tropical regions removal

of the natural forest cover has yielded infertile soils, susceptible to erosion and of very low agricultural value. In these latter circumstances a fuller understanding of the controlling environmental factors could have greatly enhanced agricultural productivity without destroying the natural balance of the ecosystems.

Further information on world vegetation may be found in Eyre (1968), Odum (1971), Polunin (1960) and Walter (1973).

4.1   Tundra

Tundra (from the Finnish word for an open, treeless plain) occurs beyond the northernmost limit of trees in northern Eurasia, northern North America, Greenland, Iceland and on other polar islands. The climate of the tundra is dominated by long, dry, severely cold winters. Annual precipitation is low (often less than 250 mm) and the air temperature for three-quarters of the year never rises above freezing, and, on occasion, may fall below −50°C. In addition, the sun barely rises above the horizon for several months of the winter and during this time of continuous darkness there is significant heat loss from the land. The ground is frozen and covered in ice and snow to a variable depth and strong winds blow the snow into tightly packed drifts. Tundra summers are cool and short and, despite the long daily duration of sunlight, less than three months have an average temperature above freezing and the mean temperature of the warmest month does not exceed 10°C. Except for a few centimetres at the surface which are subject to summer thawing (the active layer) the ground remains permanently frozen, often to a great depth (permafrost).

Tundra soils are susceptible to frost churning and, therefore, rarely develop strong zonal characteristics. They range from thin, stony lithosols and regosols to gley soils and waterlogged peats where free drainage is impeded by an impervious substrate (either rock or permafrost). The combination of permafrost and the repeated seasonal thawing and freezing of the active layer makes the soil very unstable. Solifluction (the movement of soil and rock fragments down slopes) is a common feature resulting from the movement of water-saturated soil over the lower, frozen permafrost layer. In addition, stone and soil polygons and circles are common features of flatter areas: this 'patterned ground' is produced by the differential movement of frost-churned particles.

As a result of the harsh environment, only a few well-adapted plants can survive in the tundra. The short growing season necessitates a rapid seasonal life-cycle, so that annual species are rare. Perennials, which can commence growth early from buds formed at the end of the previous season, are favoured. This rapid response to moderating climatic conditions is facilitated by food reserves stored in modified prostrate stems, rhizomes, bulbs and corms. Reproduction is mainly vegetative, by shoots, runners and bulbils. The frozen soils, low annual precipitation, frequent strong winds and low relative humidity combine to limit the amount of water available for plant growth, so that many species have xeromorphic adaptations to their leaves and stems, and grow in compact, low-growing mats, cushions and rosettes for maximum protection.

The vegetation of the dry, northerly regions of the tundra is a tundra heath, dominated by lichens (in particular, species of *Cladonia, Cetraria* and *Alectoria*) and mosses. Further south, or where there is more available soil moisture, a tundra grassland or heathy moorland develops, dominated by *Carex* spp. (sedges) and *Eriophorum* spp. (cotton-grasses) of the Cyperaceae. On the tundra moorlands dwarfed members of the Ericaceae, including *Empetrum nigrum* (crowberry), *Cassiope* spp. (arctic bell-heather) and *Vaccinium* spp. (bilberry and cowberry) assume dominance. On south-facing slopes, in damp depressions and by watercourses, dwarf shrub thickets of *Salix* spp. (willows), *Betula* spp. (birches), *Juniperus* spp. (juniper) and, locally, *Alnus* spp. (alders) predominate. In the grasslands and heathlands several attractive flowering species of *Ranunculus* (buttercup), *Dryas* (avens), *Saxifraga* (saxifrage) and *Gentiana* (gentian) can often be found, whilst where drainage is impeded, bogs and swamps containing species of *Sphagnum* (bog moss) and *Eriophorum scheuchzeri* (Scheuchzer's cotton-grass) are common.

There is no equivalent of Arctic tundra in the southern hemisphere since the Antarctic land mass has an almost complete cover of ice. On the Antarctic continent itself only two species of vascular plant, the hair-grass *Deschampsia antarctica* and *Colobanthus quitensis*, a member of the Caryophyllaceae (the Pink family), are known to grow. The land plant life, therefore, consists almost entirely of non-vascular cryptogams (bryophytes and lichens), although even the distribution of these is limited − not so much by low temperature, but rather by the excessively dry conditions, since most of the soil water is frozen. On some sub-antarctic islands, a tundra-like vegetation with a greater diversity of vascular plants occurs, including members of the genus *Azorella* of the Umbelliferae in association with *Acaena* spp. of the Rosaceae and various grasses, mosses and lichens.

4.2 Taiga

Taiga (from the Russian word for a marshy, coniferous forest) forms an enormous, unbroken vegetation zone throughout the northern parts of North America, Europe and Asia. The northern boundary of these coniferous forests forms the Arctic tree-line, beyond which lies tundra. However, the southern boundary with the temperate forests is less distinct. There is no equivalent of taiga in the southern hemisphere, where there is no land at the appropriate latitudes.

The climate of the taiga is 'cold temperate'. During

4.3 (a) View of Arctic tundra on Spitzbergen. Plants are colonising a glacial moraine. (b) *Eriophorum scheuchzeri*, a tundra species of wet peaty soils.

4.4 (a) View of Antarctic tundra with ice-covered land in the background. (b) Antarctic tundra vegetation, mainly low-growing grass and non-vascular cryptogams (bryophytes and lichens).

the long, cold months of winter, temperatures fall well below freezing, especially in continental interiors, and the frozen ground has a thick covering of snow. Summers are short, but often quite warm, and the mean temperature of the warmest month is always above 10°C. Annual precipitation is moderate (from 250 mm inland to 1000 mm on western seaboards), but evaporation is limited for most of the year by the low temperatures, so that humidity remains high. The typical soils of these regions are podzols or, in areas of impeded drainage, peat.

As a result of their ability to survive a relatively short growing season, the taiga is dominated by coniferous trees which, apart from species of *Larix* (larch), retain their leaves throughout the winter months. Leaves of coniferous trees are reduced to xeromorphic needles or scales, so that during the winter, when water uptake from the frozen ground is impossible, transpirational losses are kept to a minimum. The trees are shallow-rooted, above the permanently frozen subsoil. Intermingled with the relatively pure stands of coniferous species are a few hardy broad-leaved deciduous trees, although their occurrence, except at the northern boundary of the taiga, often follows the removal of the original trees by burning or felling. They include members of the genera *Betula* (birches), *Populus* (aspens and poplars), *Alnus* (alders) and *Sorbus aucuparia* (rowan).

Throughout western Europe *Pinus sylvestris* (Scot's pine) and *Picea abies* (Norway spruce) dominate the taiga zone. These are gradually replaced eastwards in European Russia and in Asia by *Picea obovata* (Siberian spruce), *Abies sibirica* (Siberian fir), *Larix sibirica* (Siberian larch) and *Pinus cembra* var. *sibirica* (Siberian stone-pine). In central and eastern Siberia, where the permafrost lies close to the surface, the shallow-rooting *Larix gmelinii* (Dahurian larch) and *Pinus pumila* (Siberian dwarf-pine) increase in abundance. The ultimate dominants of the eastern Asian taiga are *Abies veitchii A. sachaliensis* and *Picea ajanensis* in association with a great variety of subsidiary tree species.

The flora of the North American taiga is equally varied. In the east *Picea glauca* (white spruce) and *Abies balsamea* (balsam fir) grow on well-drained soils, while on poorer, wetter soils *Picea mariana* (black spruce), *Larix laricina* (tamarack) and *Pinus banksiana* (jack pine) predominate. Westwards these are replaced by *Pinus contorta* (lodgepole pine) and *Abies lasiocarpa* (alpine fir).

As a result of the dense shade cast by the coniferous canopy and the deep, acid pine needle litter on the ground, the shrub and herb layers of all taiga forests are poorly developed, usually consisting of *Vaccinium* spp. (bilberry), *Ledum* spp. (Labrador tea) and *Empetrum nigrum* (crowberry). In damp depressions the ground is covered with feather-mosses (e.g. *Hylocomium* spp.), but further north, and on drier areas, lichens such as *Cetraria* spp. (Iceland moss) and *Cladonia* spp. (reindeer moss) form an unbroken carpet.

Near the northern limit of the taiga is an extensive area of forest tundra, characterised by scattered, stunted coniferous trees or deciduous scrub, predominantly of *Betula* spp., which marks the transition zone between taiga and tundra. To the south, the taiga is bounded by a transitional zone of mixed coniferous and deciduous trees, which gives way to the temperate forests. In western and central Europe these mixed forests contain *Pinus sylvestris* and *Picea abies* in addition to deciduous trees, whilst in North America *Pinus strobus* (white pine), *P. resinosa* (red pine) and *Tsuga canadensis* (eastern hemlock) are the principal conifers.

4.3 Temperate forests

4.3.1. *Broad-leaved deciduous forests*

Forests of this type once covered most of the lowlands of eastern North America, western and central Europe and parts of China and Japan. However, they are now much reduced in extent as a result of clearance for cultivation, grazing land, fuel, settlement and timber.

The climate of these forests is cool temperate. The cold winter season, during which up to four months may have a monthly mean temperature below freezing, are followed by the warmer spring and summer, when at least six months have a mean temperature above 10°C. In continental regions the annual rainfall is about 500 mm and shows a distinct summer maximum. These climatic features are moderated in areas where there is an oceanic influence. Along the western seaboards of Europe, for example, the rainfall is high (in excess of 800 mm per annum and often far higher) and is more evenly distributed throughout the year than in continental regions. Temperatures are also more equable and seasonal fluctuations less well marked.

The most widespread and characteristic soil of this formation type is the brown earth but in high rainfall districts and on sands and gravels, podzols, or in extreme cases, peats may form.

Temperate forests are all structurally similar, with a

single tree layer (often dominated by more than one species), a shrub and sapling layer, a herb layer (frequently containing spring perennials with underground storage organs) and a ground layer of mosses and lichens. Climbing and epiphytic angiosperms are few. The shrub layer may assume dominance following removal of the main canopy or at the forest edge. The trees are typically broad-leaved and deciduous and the annual leaf fall is a response to the lower temperatures of winter, when transpirational losses from leaves would otherwise exceed water uptake from the cold or frozen soil. Most of the associated herbs and shrubs of these forests die back during the winter months and, as a consequence, the seasonal aspects are particularly well marked.

In western and central Europe the major forest-forming deciduous trees are *Quercus petraea* (sessile oak), *Q. robur* (pedunculate oak) and *Fagus sylvatica* (beech), which may be locally dominant or form mixed forests with *Ulmus* spp. (elms), *Fraxinus excelsior* (ash), *Tilia* spp. (limes), *Quercus cerris* (turkey oak) or *Carpinus betulus* (hornbeam). The latter two species have a more southerly distribution than the others. Further east, other species of *Quercus* and *Parrotia persica* (Persian ironwood) increase in importance. Locally common trees include *Acer* spp. (maples), *Betula* spp. (birches) and *Castanea sativa* (sweet chestnut), whilst *Alnus glutinosa* (alder), *Populus* spp. (poplars) and *Salix* spp. (willows) dominate in damper situations. The degree of development of the shrub and herb layers of these woodlands is variable: in beech forests it is minimal as a result of the deep shade cast by the closed tree canopy and the deep litter layer of acid beech leaves, but in more open canopy woodlands the species diversity increases to include many shrubs and small trees such as *Corylus avellana* (hazel), *Prunus* spp. (cherries) and several other shrubby members, of the Rosaceae. The commonest climbers are *Hedera helix* (ivy) and *Lonicera periclymenum* (honeysuckle). The herb layer comprises many flowering plants, grasses and some ferns; in damp areas or high rainfall districts the bryophyte layer is well developed. Broad-leaf deciduous forests of the British Isles are described in detail in section 5.2.

Much of western and northern Japan and eastern China support mixed deciduous forest. In the north the dominant species include *Quercus mongolica* (Japanese oak), *Acer japonicum* and *A. sieboldiana* (maples), *Tilia* spp. (limes), *Betula* spp. (birches) and *Ulmus* spp. (elms) with some *Pinus koraiensis* (Korean pine). These trees are augmented further south by many other deciduous and coniferous species, including *Liriodendron chinense* (Chinese tulip tree), *Fagus crenata* and *F. japonica* (beech), *Carpinus* spp. (hornbeams), *Cryptomeria japonica* (Japanese cedar) and other conifers, plus creepers such as *Vitex* spp. (giant wild vines).

In North America, the transition forests to the south of the taiga give way east and south of the Great Lakes to a broad-leaved forest of *Fagus grandiflora* (American beech), *Acer saccharum* (sugar maple), *Tilia americana* (basswood) and *Quercus* spp. (oaks). In the southern Appalachians these forests are replaced by the more luxuriant 'cove hardwood' forests, which are dominated by varying mixtures of *F. grandiflora*, *Quercus rubra* (red oak), *Q. alba* (white oak), *Tilia* spp. (basswoods), *Acer saccharum*, *Tsuga canadensis* (hemlock) and *Liriodendron tulipifera* (tulip tree), which has a southerly distribution. To the south, west and east of these mixed hardwood forests, a further type of deciduous woodland which includes *Carya* spp. (hickory) occurs.

In South America, deciduous forests are found in the very south of Chile and southern Tierra del Fuego. The dominant trees are deciduous species of *Nothofagus* (southern beech) in association with evergreens, of which *Drimys winteri* (Winter's bark) is an example.

4.3.2 *Coniferous forests*

These forests are only found in cool, temperate regions in which the rainfall is very high. The largest and finest example is the Pacific 'coast forest' of North America, which lies to the west of the coastal mountains between southern British Columbia and northern California. Frequent cyclonic storms from the Pacific Ocean moderate winter and summer extremes of temperature and give rise to the very high rainfall (between 800 and 3800 mm annually), most of which falls in winter. These high levels of precipitation, combined with the cool temperatures, contribute to a high atmospheric humidity. Soils are podzolic with a deep surface layer of humus.

The dense coastal forests of North America support some of the world's largest trees, including *Tsuga heterophylla* (western hemlock), *Thuja plicata* (western red cedar), *Pseudotsuga menziesii* (douglas fir), *Picea sitchensis* (sitka spruce) and *Sequoiadendron giganteum* (the big tree or wellingtonia), in addition to *Pinus* spp. (pines), *Larix* spp. (larches) and other species of *Pseudotsuga* and *Picea*. Southwards, the large *Sequoia sempervirens* (Californian redwood) increases in dominance. These trees, which often exceed heights of 100 m

a

b

4.5   Pacific coast forest: (a) *Picea sitchensis* in the Queen Charlotte Islands, Canada. (b) *Thuja plicata* and *Tsuga heterophylla* forest on the west coast of British Columbia, Canada.

and trunk girths of 20 m, render the forest floor so dark and the humus layer so deep that only ferns, mosses and a few shade-loving shrubs and herbs can survive beneath them.

4.3.3   *Broad-leaved evergreen forests*

Examples of this forest are found in south-eastern North America, north-east Mexico, southern Japan, central Chile, parts of New Zealand and Tasmania, and the extreme tip of South America, all of which have a warm, but wet, temperate climate.

The precipitation in these regions is high (usually in excess of 1500 mm annually, but this may be doubled on exposed hills and coasts). The summers are long, warm and humid and there is frequently a summer rain-

fall maximum. Winters are mild, with rain more common than snow, and frosts only occur occasionally. Coastal zones may be affected by typhoons.

The soils of these forests are usually podzolic as a result of the high rainfall. A further consequence of the wet climate, and associated high atmospheric humidity, is that these forests often drip with moisture, giving them certain similarities with tropical rain forests (hence the epithets 'dripping' or 'temperate rain' forests). The dominant tree species may support several climbing and epiphytic plants, and where the canopy is less dense, a subsidiary tree stratum of palms and tree-ferns may be present. The herb and shrub layers are dense.

In North America broad-leaf, evergreen forests occur sporadically in the southern states which border the Gulf of Mexico (but excluding southern Florida) and these are dominated by several species of evergreen oak, in particular *Quercus virginiana* (live oak), in addition to *Magnolia grandiflora* (evergreen magnolia). The trees support a variety of climbers and are festooned with epiphytic herbaceous species, including the bromeliad *Tillandsia usneoides* (Spanish moss). In the frequent swamps *Taxodium distichum* (swamp cypress) predominates; on sandy soils, species of pine are prominent, with *Pinus palustris* (longleaf), *P. caribaea* (slash) and *P. taeda* (loblolly) amongst the commonest.

In southern Japan and central China these forests are composed of a large variety of trees – principally evergreen *Quercus* spp., but also *Lithocarpus* spp. (tan oak), *Castanopsis* spp. (chinquapin) and *Castanea* spp. (sweet chestnut) in addition to numerous evergreen members of the *Lauraceae* (the laurel family), *Magnoliaceae*, *Theaceae* (the tea family) and *Hamamelidaceae* (witch-hazel family). Shrubs and bamboos form a dense undergrowth and woody climbers, epiphytic ferns and orchids are plentiful.

The Australasian forests of this type are found in the super-humid regions of Tasmania and New Zealand. They are characterised by large evergreen species of *Nothofagus* (southern beech) in addition to giant conifers such as *Agathis australis* (kauri pine), *Dacrydium cupressinum* (red pine), *D. franklinii* (huon pine) and species of *Podocarpus* and *Araucaria*. Magnificent groves of tree-ferns, which may attain a height of 10 m, frequently form an understorey and the well-developed shrub, herb and ground layers contain a great variety of climbers, epiphytes, ferns and bryophytes.

The Chilean temperate rain forest occupies the south-western part of the country. Several evergreen *Nothofagus* spp. (southern beech) and conifers are the principal trees, and, as above, other forest strata are also well represented. Thickets of *Chusquea* spp. (wild bamboo) contribute to the impenetrable undergrowth of these forests.

The temperate forests in the extreme south-east of South Africa have been heavily exploited for timber. The small remnants contain tall coniferous *Podocarpus* spp. (Yellow-woods), which often grow to a height of 45 m, in association with *Myrsine melanophloeos* (Cape beech), *Olea laurifolia* (black ironwood) and several other evergreen broad-leaf species.

4.3.4 *Evergreen mixed coniferous and sclerophyllous forests*

Forests of this type were once typical of regions with a 'mediterranean' climate, but, as a result of man's activities, particularly the grazing of domestic animals, most have disappeared and been replaced by low shrub vegetation. These areas include much of the European Mediterranean region, the extreme south-west of North America and South Africa, parts of southern Australia and adjacent Tasmania, and central Chile.

Hot, dry summers and mild winters with some rain typify the mediterranean climate. Precipitation is low (between 250 and 1000 mm annually) and falls mainly during the winter; mid-winter snows and frosts occur infrequently. Soils are varied. Thin rendzinas and terra rossa (red soils) have developed on limestones. The latter are residual soils rich in red iron oxides that accumulate in depressions under conditions of prolonged summer drought. On siliceous material, various brown soils (terra fusca), which are deeper and have a higher organic content than the limestone soils, occur.

As a result of the marked summer drought the vegetation is xeromorphic in nature. Trees and shrubs possess small, leathery (sclerophyllous) or needle-shaped, evergreen leaves and a thick bark, which restrict transpirational water loss. Some species have deep root systems which reach the low summer water table, whilst succulent herbaceous species store water in specialised stems and leaves to tide them over the dry season. Other plants flower very early in the year during the damp winter and spring months, their growth facilitated by the mobilisation of food reserves stored in underground bulbs and tubers.

The lands bordering the Mediterranean and southern Black Sea once supported an open, sclerophyllous woodland of evergreen oaks, which included *Quercus suber* (cork oak), *Q. ilex* (holm oak) and *Q. coccifera* (Kermes oak), with the Pines *Pinus pinea* (stone-pine), *P. halepensis* (Aleppo pine), *P. nigra* var. *maritima* (Corsican pine) and *P. pinaster* (maritime pine), together with *Cedrus libani* (Cedar of Lebanon). Most of this woodland on the richer terra fusca soils has long since been felled and replaced by a sclerophyllous scrub vegetation known as 'maquis', and on the drier limestone soils or in areas subject to intensive burning and grazing by a low-growing, often spiny vegetation known as 'garrigue'. These communities vary from tall, dense shrubby thickets in which a few, small trees remain, to low, treeless scrub. The shrubs of the maquis are extremely profuse and include, for example, *Olea europea* (wild olive), *Pistacia lentiscus* (mastic), *Cistus* spp. (rock roses), *Myrtus communis* (myrtle), *Nerium oleander*

(oleander), *Prunus* spp. (laurels), *Genista* and *Ulex* spp. (brooms and gorses), *Juniperus* spp. (junipers), *Erica* spp. (tree-heaths) and *Arbutus* spp. (strawberry trees). In addition, there is a great diversity of xerophilous grasses, succulents and colourful, spring-flowering herbs. Garrigue also supports several aromatic herbs, including *Thymus* spp. (thyme), *Lavandula stoechas* (lavender) and *Rosmarinus officinalis* (rosemary).

In California and some adjacent regions the dominant trees are small evergreen oaks, including *Quercus agrifolia* (California live-oak) and the shrub-like *Q. dumosa* (California scrub-oak), with several coniferous species. Many of these woodlands have been removed and replaced by 'chaparral', a tall, dense, maquis-type vegetation, composed of a great variety of shrubs and herbs.

In South Africa the 'Cape maquis' is confined to the south-west and south of Cape Province, and is now devoid of trees. The tall shrubs belong to various families, in particular the Proteaceae (the silk oak family), the Restionaceae (sedge-like plants) and the Ericaceae (tree-heaths).

In Australasia, sclerophyllous vegetation has developed in northern and central Tasmania and south-west, south and south-east Australia. The forests are dominated by giant *Eucalyptus* spp. (gum trees), which can attain an average height in excess of 70 m. In the undergrowth species of *Acacia* (wattles), *Mimosa* and heath-like members of the Epacridaceae are abundant. These forests have been extensively exploited for timber and, in areas of disturbance or lower rainfall, the characteristic vegetation is either dwarf *Eucalyptus* scrub, known as 'mallee', or dwarf *Acacia* scrub, known as 'mulga', with occasional stunted gum trees and evergreen trees of *Casuarina stricta* (she-oak), the leaves of which are reduced to scales.

The Mediterranean zone of central Chile supports a coastal maquis-like vegetation which includes shrubby members of the Solanaceae (the potato family), and small trees, such as *Acacia*. This gives way inland at higher altitudes to a sclerophyllous woodland where characteristic evergreen plants are *Quillaja saponaria* (Chilean soapbark), *Escallonia* spp. (Chile gumbox), the coniferous *Araucaria araucaria* (monkey puzzle), and species of *Crinodendron*.

4.3.5   *Montane forests*

On mountain slopes throughout the temperate zone, forests dominated by coniferous species replace the lowland formations. These montane forests occur on the mountain ranges of North America, Europe and Asia and, to a lesser extent, in the southern hemisphere in South America and New Zealand. They extend to the tree-line, the upward limit of tree growth. The altitude of the tree-line varies with latitude and is frequently marked by a zone of 'krummholz' (German: gnarled wood), in which the trees are dwarfed and deformed by the rigours of the environment. Beyond this limit, on the higher mountains, montane forest is replaced by alpine vegetation (sections 4.9 and 9.4). At their lower limits a transition zone of mixed coniferous/deciduous forest may border the lowland forest, but nearer the tree-line and on more northerly mountain ranges these forests are very similar to taiga.

At the lower limits of the montane zone the climate is little different from that of the regional formation. However, with increasing altitude the conditions become cooler and wetter. Wind speed and the depth and duration of winter snow cover also increase (section 9.1). These forests occupy the zone of maximum precipitation on mountain slopes, with an annual total usually in excess of 1000 mm, which gives rise to soils that are often podzolic.

In the mountains of Europe the principal trees are *Picea abies* (Norway spruce) and *Abies alba* (silver fir), although *Larix decidua* (European larch) and *Pinus mugo* (mountain pine) and *P. cembra* (Arolla pine) increase in abundance towards the tree-line. The shrub and herb layers of these forests are not well developed owing to low light levels under the canopy and the thin, rocky, acidic soils. However, a rich ground layer of bryophytes and lichens is a common feature.

On the west coast mountains of North America the dominant trees are coniferous, although in sheltered positions deciduous hardwoods such as *Populus tremula* (aspen) and *Betula* spp. (birches) may flourish. The lower montane forests contain several species of pine, including *Pinus ponderosa* (ponderosa pine), *Pseudotsuga menziesii* (douglas fir), *Abies concolor* (white fir) and *A. grandis* (grand fir). The higher, sub-alpine forests contain *Picea engelmannii* (Engelman spruce) together with other species of fir and pine, in particular *Abies lasiocarpa* (alpine fir) and *Pinus contorta* var. *latifolia* (lodgepole pine).

4.4   Temperate grasslands

Grasslands occur throughout the temperate zone wherever rainfall is too low and/or periods of drought are too long and frequent to support forest growth.

4.6 (a) Tree-line in the Italian Dolomites, Becco di Mezzodi, 2602 m, in the background. (b) European alpine coniferous forest (*Pinus mugo* growing in the Italian Dolomites).

Appropriate climatic conditions prevail in the interiors of large continents and have given rise to the prairies of central and western North America, the steppes of eastern Europe and central Asia, the pampas of Argentina, the South African veldt and the tussock grasslands of parts of New Zealand and eastern Australia. These naturally occurring grasslands should not be confused with pastures in the temperate forest zone which are maintained by grazing or mowing.

The climate of this zone is one of seasonal extremes: cold, dry winters, in which temperatures often fall well below freezing, give way to the warm, moist months of spring and early summer when the rainfall maximum usually occurs. The later, summer months are dry and very hot. Annual precipitation is low (between 250 and 750 mm), and evapotranspiration is high on account of the strong, drying winds. Winters are less extreme in the southern hemisphere grasslands than in those of the north.

The typical soils of these grasslands are unleached chernozems (black earths), which are extremely fertile with a deep, dark, humus-rich horizon (up to 1.5 m thick) directly overlying the calcium-rich substratum. These soils are replaced in warmer, drier regions by chestnut soils, which are lighter in colour, and in wetter areas by prairie soils (leached chernozems). The intense heat of late summer often scorches the dry vegetation, causing grassland fires which enrich the soil with ash and inhibit tree growth.

As a result of their fertility and the favourable climate most natural grasslands have been replaced by intensive cereal farming, although in drier regions, where the land use is mainly pastoral, some of the original vegetation cover still remains. The dominant plants of these grazed areas are grasses, but in addition there are low, woody plants and numerous herbs (principally members of the Compositae and Fabaceae) which add colour when they flower in the spring and autumn. The height of the grasses ranges from over 2 m in the more humid areas, to a few centimetres in the driest parts.

The prairie grasslands of central North America extend from southern Alberta in Canada to Mexico and from the arid deserts of the west to the more humid eastern forests. The moist eastern grasslands, although now mostly removed for agriculture, support a dense growth of tall 'bunch grasses' – species such as *Andropogon* spp. (bluestems) and *Sorghastrum nutans* (Indian grass) – in addition to *Stipa* spp. (feather-grasses) and *Panicum vergatum* (switch-grass). Trees and shrubs

occur at the forest/prairie boundary, but further west these only occur along watercourses. As rainfall decreases westwards, the bunch grasses become shorter and less dense and the 'tall-grass' prairie grades into 'mid-grass' and eventually 'short-grass' prairie. The principal species here are *Bouteloua* spp. (gramas), *Agropyron* spp. (wheat-grasses), *Buchloe dactyloides* (buffalo grass), *Stipa* spp. (feather-grasses) and xeromorphic shrubs such as *Prosopis* spp. (mesquite).

The steppes form an almost continuous band eastwards from central Hungary (where they are called pusztas) through southern Russia and northern Mongolia to north-eastern China. The transition zone between these grasslands and the temperate forests further north is called 'forest steppe'. Rather than a gradual decrease in tree cover, this zone consists of alternate patches of deciduous forest and grassland. Further south, except for occasional small riverside groves, trees are totally absent from the steppes since scarcity of water, severely cold winters and strong winds prevent their growth. During this century, however, some drought-resistant trees have been planted as windbreaks across the steppes. Characteristic steppe species are *Stipa* spp. (feather-grasses), *Festuca* spp. (fescues) and *Koeleria gracilis* (hair-grass), all of which possess a well-developed, deep, fibrous root system and xeromorphic above-ground features to minimise water loss. In addition to low-growing woody plants such as *Artemisia* spp. (wormwood) flowering plants with underground perennating organs predominate. As in North America, little of the Eurasian temperate grasslands has survived; they have been replaced by cereal crops or pasture.

In South America, pampas grasslands are characteristic of the vast plains of north-eastern Argentina and adjacent regions of Uruguay and southern Brazil. The humid pampas in the east, which are now largely cultivated or grazed, naturally support tall feather-grasses with a few shrubs along watercourses. However, in the sixteenth century Spanish settlers introduced European grasses which, as a result of their dense turf-forming habit, replaced most of the native grasses.

In South Africa grasslands occupy the higher plateaux and low mountain slopes in the east of the country. The dominant species is *Themeda triandra* (red grass), but retrogression to coarser, xerophytic grasses (e.g. *Eragrostis* spp. (love grass), *Aristida junciformis* (bristle grass) and *Cynodon dactylon* (Bermuda grass)) has often followed overgrazing and soil erosion.

Tussock grasslands are the natural vegetation cover

4.7 In the rain shadow of the Rocky Mountains, dry, heavily grazed steppe near Lake Kamloops, British Columbia.

of the eastern half of South Island, New Zealand, but these are now nearly all cultivated or grazed. The most conspicuous species are *Danthonia* spp. (poverty grass), *Festuca novae-zealandiae* (New Zealand fescue) and *Poa cespitosa* (tussock grass). As in other grassland areas, soil erosion has often followed excessive grazing.

4.5 Tropical forests

4.5.1 Rain forests

Species-rich, broad-leaved, evergreen rain forests are characteristic of equatorial lowlands where the climate remains warm and wet throughout the year. The main areas of this forest are:

(a) on the American continent: in the basins of the Amazon and Orinoco rivers, along the east coast of Brazil, and along the Atlantic coast of central America;

(b) on the African continent: in the Congo basin, along the south-west Atlantic coast and on the east coast of Madagascar;

(c) in western India, Ceylon, Burma, the Philippines, Indonesia and much of the Indo-China peninsula and the Malay archipelago; and

(d) along the north-east coast of Australia.

A distinctive feature of the climate of these regions is the absence of either a dry or a cold season. The annual rainfall is high (usually in the range 2000–4000 mm, but occasionally exceeding 10 000 mm) and evenly distributed throughout the year. There is little annual temperature variation and the yearly mean lies around 26–27°C. Inside the forests humidity is always high.

The forest soils are red or yellow-brown latosols which have developed under the constantly wet and warm conditions. Weathering of the parent material has proceeded to a great depth (more than 20 m in some places) under highly oxidising conditions and has been accompanied by strong leaching, which removes plant nutrient elements and silicates from the upper soil layers. This has produced soil profiles which are uniform in structure and rich in the products of weathering, namely hydrated iron, aluminium and titanium oxides. This is the 'laterisation' process. Latosols are almost entirely mineral, with little or no humus owing to the rapid decomposition of organic material. They are nutrient-deficient, and are weakly to strongly acidic in reaction. The paucity of nutrients is paradoxical in view of the abundance and diversity of species that these soils support. However, rapid decomposition of organic material ensures that the surface soil is constantly enriched and any small losses to the system in drainage water are quickly replaced by weathering of the underlying rock. Human interference has a drastic effect on these forests; in particular, tree felling and burning greatly diminish

4.8   Tropical rain forest which has been logged and burned prior to land conversion to agriculture, Malaysia

the pool of available nutrients so that subsequent plant productivity is reduced.

The tropical rain forest vegetation is extremely diverse with, on occasion, more than 250 tree species per hectare. Nearly all species of woody plant in these forests are evergreen, which enables them to take advantage of the continuous growing season. They have large, entire, dark-green, leathery leaves which are often extended into 'drip tips', from which the rain flows easily. The trees form a dense forest within which at least three strata can be distinguished. The main canopy layer is dense and at an average height of 30 m. Above this scattered emergent trees may extend to 60 m or more, and below, smaller, slender trees of the sub-canopy form a broken layer between 5 m and 15 m in height. The trunks of the canopy trees are very straight and many have 'plank buttresses' at their bases (Richards, 1952).

Beneath the trees, light intensity is reduced to as little as 1% of that above the canopy and direct sunshine only reaches the forest floor as small 'sun flecks'. Consequently, except along the edges of clearings and watercourses, the shrub and herb layers of the forests are poorly developed and there is much bare ground. The thick-stemmed climbing 'vines' (lianas), which root on the forest floor and bear their leaves and flowers high up in the canopy, and woody and herbaceous epiphytes of tree trunks, branches and leaves, are abundant. Epiphytic ferns, orchids, bromeliads, bryophytes and lichens use the trees for support. Most grow in a soilless environment, but some have aerial roots that trap leaf litter, from which mineral nutrients can be extracted. This trapped humus also acts as a sponge to collect and hold water, although in other species urn-shaped leaves perform this function. 'Stranglers' are canopy epiphytes which later in their life-cycle send down long, pendant roots to the soil. These roots gradually increase in number and thickness until the host tree is totally encased by them. Eventually, the host tree dies and decomposes, leaving a hollow structure of independently growing vine in its place.

Saprophytic plants are an important component of the forest ecosystem. These plants are devoid of photosynthetic tissues and their nutritional requirements are supplied instead from decomposing organic material. The majority of saprophytes are fungi and bacteria, although there are also some small saprophytic orchids, gentians and other higher plants. Parasitic species are also abundant and of two types: terrestrial root parasites and epiphytic semi-parasites. There are only two families of root-parasites, with few members in each – the Balanophoraceae and the Rafflesiaceae. The semi-parasites all belong to the Loranthaceae (mistletoe family), members of which are numerous and abundant on a variety of hosts throughout these forests.

The largest area of rain forest in the Americas is in

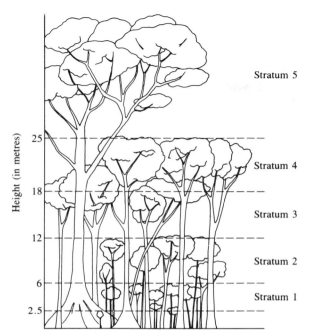

Height (in metres)

25
18
12
6
2.5

Stratum 5
Stratum 4
Stratum 3
Stratum 2
Stratum 1

4.9 Diagrammatic section through tropical rain forest vegetation from the Ivory Coast, West Africa. The dotted lines show the limits of five strata within the forest which can be recognised (modified from Bourgeron & Guillaumet, 1982).

the Amazon basin, where trees of the Fabaceae (the pea family) such as *Dalbergia* spp. (rosewood) and *Haematoxylon brasiletto* (Brazilwood) are the most abundant, but members of the Lauraceae (laurels), the Rosaceae, Burseraceae, Meliaceae (mahoganies), Rubiaceae (the coffee family) and the Sapotaceae (which includes *Achras sapota* (sapodilla) from which chicle, the base for chewing gum is derived), to mention but a few, are also well represented. Palms are also abundant, as are epiphytic orchids. By watercourses more light penetrates to lower levels and the vegetation is correspondingly denser, and species such as *Hevea braziliensis* (para rubber tree) and *Ceiba* spp. of the Bombacaceae, from which kapok is obtained, can be found. Smaller plants include species of *Clusia* (strangler vines) and herbs, principally of the Marantaceae (the arrowroot family) and of the Melastomataceae. On the northern South American coast, in the West Indies and in Central America other tree species, e.g. *Mora* spp., *Guiacum officinale* (lignum vitae) and *Ocotea* spp. (greenheart), increase in dominance. Tree-ferns are abundant in some areas.

The African rain forests, when compared to those of South America and Asia, are relatively impoverished in species. However, many of the larger trees yield valuable timber, for example, *Khaya* spp. (African mahogany), *Lophira* spp. (African oak), *Triplochiton* spp. (obeche) and *Coula edulis* (African walnut). Large members of the Fabaceae (e.g. *Brachystegia* and *Berlinia* spp.) are also important. Sub-canopy trees and shrubs include many members of the Rubiaceae, Euphorbiaceae and Sterculiaceae (e.g. *Cola* spp. (wild kola nut)) and *Diospyros* spp. (ebonies) of the Ebenaceae. Lianas, such as the rubber-producing *Landolphia*, and epiphytes, including many ferns, begonias and orchids, abound. Palms, e.g. *Raphia* spp., and bamboos are also characteristic, especially in wetter parts.

The Indian-Malayan rain forests are, floristically, the richest of the formation. There is an enormous variety of trees, many of which belong to the Dipterocarpaceae e.g. *Dipterocarpus* and *Hopea*, but, in addition, there are representatives of the Fabaceae, Annonanaceae (custard apple family) and other families. Palms, including *Cocos nucifera* (coconut palm) also occur. Epiphytic plants, mainly orchids and ferns, form a prominent component of the forest – in the Philippines alone 900 members of the Orchidaceae occur. A large variety of lianas are also present, including the rattans (*Calamus* spp. (climbing palms)), which climb by means of thorns on their leaf tips. The largest known flowers in the world are found on species of *Rafflesia*, which are parasitic on certain lianas in the Malayan forest.

In regions of the humid tropics where the precipitation is more seasonal (i.e. up to three months in the year may receive only 5–10 cm of rain) the rain forest is replaced by evergreen seasonal forest. This is considered to be a sub-formation of the rain forest, which it closely resembles, although the vegetation is lower growing and the upper tree stratum (the emergent stratum) is discontinuous. These forests are predominantly evergreen, but up to 25% of the species present, including many of the larger trees, may be deciduous during the dry season.

On sheltered, muddy coasts, in salt marshes, and along brackish estuaries in both tropical and subtropical regions, the typical vegetation is dense 'mangrove swamp-forest'. This reaches its best expression in regions where tropical rain forest is the climax vegetation away from the coast. The small evergreen trees (up to 30 m in height) and shrubs of the swamp, which are members of various families, all produce descending adventitious roots which form an arch from the trunk of the mangrove

a

b

4.10 Mangrove swamp forest, Malaysia: (a) coastal *Avicennia* zone; (b) seed of *Rhizophora* sp. germinating on the parent plant, the radicle penetrates the mud when the young plant falls to the ground – an adaptation which prevents removal by the tide.

to reach the muddy substrate at some distance from the parent plant. These aerial 'stilt roots' lift the main trunk above the water level. New trees can develop adventitiously from the point of contact of the roots and the soil, so that the forest spreads rapidly. Mangrove trees also produce 'breathing roots' (pneumatophores) which project upwards through the mud; air diffuses down these structures into the submerged portions of the plant. In addition, the roots of mangroves trap silt and mud, which raises the land surface and the forest then extends further seawards, a feature that is of economic importance in reclaiming land from the sea in the tropics.

The mangrove forests of tropical America are dominated by *Rhizophora mangle* (common or red mangrove) of the Rhizophoraceae and *Avicennia nitida* (black mangrove) of the Verbenaceae, in addition to some halophytic palms. Further inland a community dominated by *Conocarpus erectus* (buttonwood) occurs. On parts of the west African coast there are analogous communities dominated by *Rhizophora* and *Avicennia* spp. in addition to members of the genus *Pandanus* (screw pines). The east coast forests, however, have a greater affinity with the mangrove formations of India and Asia where *Sonneratia* and *Bruguiera* (Lythraceae) and extremely acidic it has formed from the dead remains of forest trees and not from the activity of *Sphagnum* moss (sections 7.1.2 and 7.1.5). The peat depth varies from at least 0.5 m to as much as 20 m and has been referred to as 'ombrogenous' in nature (section 6.5), since the surface of many of these tropical lowland peatlands is raised above the water table at the present time (Anderson, 1964). However, the manner in which peat can accumulate above the water level in the humid tropics, where normally decomposition of dead organic matter proceeds with considerable rapidity, has still to be explained.

There is a well-marked zonation of types from the wetter fringes of the peat swamp forest to the drier centre, and species of *Shorea*, for example *S. albida* and *S. uliginosa* play a prominent part. Compared to the terrestrial tropical rain forest of these regions the peat swamp forest supports fewer species and some families of the former are not represented at all. Palms are few and present only in the peripheral communities of the forest, whilst in the drier centre the vegetation shows a resemblance to that of podzolised heath-forest.

Many peat swamp forests have been expoited for their timber; in addition much has been felled, burned, drained and converted to agriculture, often with disastrous consequences on account of the acidity and low nutrient status of the resultant soil.

4.5.2 *Semi-evergreen seasonal forests*

With an increase in the duration and intensity of the dry season rain forests are replaced by semi-evergreen seasonal forests, which occur along the margins of the equatorial rain forests of Africa, Central and South America, India and south-east Asia. The climate of these areas is tropical, without any marked cold season. Annual precipitation is lower than in the rain forest belt, but usually exceeds 1500 mm per annum. However, up to five months in the year may have less than 100 mm of rain. The soils of these forests are similar to those of the adjacent rain forests, although the decay of leaf litter is prevented during the dry season.

The upper, emergent tree stratum of the rain forest is absent from these semi-evergreen forests, so the canopy consists of only two tree strata. Up to 30% of the trees in the upper storey are deciduous, although many of the evergreen species in both layers may be 'facultatively' deciduous, i.e. they are capable of shedding their leaves if the dry season is particularly severe. The

4.11 Tropical peat swamp forest (Malaysia).

and *Nypa fruticans* (nipa palm) are some of the more important additional species.

In tropical lowlands in regions where the water table is high peat swamp forest has developed over marine alluvium or sediments transported from higher ground. These forests are extensive in south-east Asia and of more restricted occurrence in some Caribbean islands, the Amazon basin and the north coast of Guyana in South America. The largest concentration of peat swamp forest is in Indonesia, especially in Kalamintan and Sumatra, but there are extensive areas in Malaysia in the Malayan Peninsula, Sabah and Sarawak.

Peat swamp forests occur behind coastal mangrove swamps and extend inland where the land is barely above sea level. The peat is very different from that of temperate peatlands and, although it is low in nutrients plant diversity and luxuriance of these forests is less than in the adjacent rain forests.

Areas of semi-evergreen forest occur both to the north and south of the rain forest in Thailand, Burma and India. In Burma the dominant deciduous tree is *Xylia xylocarpa* (pyinkado), a member of the Fabaceae, which grows in association with species of *Dipterocarpus* and *Lagerstroemia*. Epiphytes (particularly orchids) and lianas are often abundant. In Africa most of the semi-evergreen forests have been destroyed by felling and burning and replaced by a savannah-type grassland, in which very few of the original tree species remain. In South and Central America, semi-evergreen forests occur locally in the West Indies, on the northern edges of the rain forest in Venezuela and Guiana, and on the southern rim of the Amazon basin in Brazil.

4.5.3 *Deciduous seasonal forests*

In tropical and subtropical areas with a very protracted dry season the semi-evergreen seasonal forests give way to deciduous seasonal forests. These are particularly well developed and widespread in those parts of Asia with a monsoonal climate and, as a consequence, are also known as 'monsoon forests'. Forests of this type are also found locally in northern Australia and on the margins of the evergreen tropical forests of Africa and Central and South America.

The seasons of the deciduous forest belt are more marked than those of the adjacent semi-evergreen forests, although temperatures are still warm throughout the year and frosts absent. The year consists of a wet season, during which most of the rain falls, and a cooler, long dry season which may last for up to six months. Annual precipitation is in the range of 1000–2000 mm and the rains are often accompanied by strong winds or, occasionally, tropical cyclones. Soils are predominantly lateritic.

The plants of these forests are better adapted to withstand a period of drought than those of the semi-evergreen forests. Within the two strata of the canopy, the trees in the upper layer are mostly deciduous and shed their leaves during the dry season, whilst those of the lower canopy are predominantly evergreen. The trees of these forests tend to have massive trunks but are more widely spaced and lower growing than those of the rain forests and the buttressing habit is not exhibited. There are few climbers and epiphytes may be entirely absent. A consequence of the annual leaf fall

is that more light reaches the forest floor and the shrub and herb layers are usually well developed. Most plants flower during the dry period, so the seasonal aspects of these forests are quite striking.

Throughout India, Burma, Thailand and other parts of south-east Asia the dominant trees of this forest are *Tectona grandis* (teak) and *Shorea robusta* (sal). In higher rainfall areas additional trees include *Xylia xylocarpa* (pyinkado), whilst in the shrub layer *Bambusa* and *Cephalostachyum* spp. (bamboos) play a dominant role. In drier areas associated trees include other *Shorea* spp., *Dipterocarpus tuberculatus* (in) and *Terminalia tomentosa*. Shrubs, including *Dendrocalamus strictus* (another bamboo), often form dense thickets below the trees. Teak and sal both provide valuable hardwood timber and, as a consequence, many of the Asian deciduous seasonal forests have been disturbed or removed.

In southern and western Central America and parts of the West Indies members of the Fabaceae comprise a large proportion of the tree species, in addition to *Ceiba* spp. (silk cotton), *Hura* spp. (sandbox) and occasional stands of palm.

4.5.4 *Montane forests*

On mountain slopes in tropical Asia, Africa, Australia, Central and South America, the lowland flora is replaced by a vegetation in which many of the species require temperate climatic conditions. These conditions occur as the climate becomes increasingly cloudy, cool and humid with increase in altitude.

In the lower montane forests the vegetation is predominantly evergreen, although the flora is less varied than that of lowland rain forest, but at higher altitudes where the winter season is more severe there is an increase in the number of deciduous species. The trees of these high-level forests are low-growing and gnarled and there is an abundant epiphytic flora of ferns and bryophytes – hence the names 'mossy forest' and 'elfin woodland' for this vegetation. As on mountains in the temperate zone, the montane forests are replaced at their upper limit by a treeless alpine zone sometimes marked by a transitional belt of 'krummholz' (section 4.3.5).

In tropical regions where the lowland climate is dry (e.g. in areas of savannah), forests may only appear in mountainous districts. In very arid areas, for example on the western slopes of the central Andes bordering

4.12 View of tropical montane rain forest, Malaysia.

the coastal desert, the forests are absent and dry grass-lands take their place.

These vegetational changes with altitude are well illustrated on the tropical mountains of East Africa, which rise up to altitudes of nearly 6000 m from dry, lowland savannah. The lower limit of the montane zone at which forest replaces grassland lies at approximately 1500–2000 m; the upper limit lies between 3000–4000 m, where upland forest gives way to tall, ericaceous scrub. The majority of the trees in the lower montane zone are evergreen and of a wide variety of genera, including *Podocarpus* (yellow-wood), *Juniperus* (cedar), *Ocotea* (camphorwood), *Brachylaena* (muhugwe), *Croton* and *Cassipourea* (pillarwood). Shrubs, climbers and epiphytes are abundant.

The mid-montane zone, which extends between about 1800–3300 m, consists mainly of bamboo woodland, which is particularly abundant towards the upper limit of this range and where the rainfall exceeds 1250 mm. *Arundinaria alpina* (mountain bamboo) is the dominant plant, forming dense thickets up to 12 m in height. Light levels on the forest floor are very low, but some climbers twist up the bamboo stems to reach the higher light intensities of the top of the canopy.

In the upper montane zone, bamboo forest is succeeded by a low-growing elphin forest (12–18 m in height) in which gnarled and crooked trees of *Hagenia abyssinica* (koso) are commonly accompanied by *Rapanea rhododendroides* and *Pinus africana* (red stink-wood) and arborescent *Hypericum* spp. This woodland zone usually occupies the wettest part of the montane zone and, as a result of the high humidity and relatively open canopy, the herbaceous layer is species-rich and bryophytes, particularly epiphytic species, are abundant. The transition to the alpine zone is marked by an ericaceous shrub belt in which large tree-heathers dominate (section 4.9).

4.6 Savannah

Savannah is a term applied to any tropical or subtropical grassland within which scattered trees and shrubs frequently occur. Some savannahs are natural, whilst on others a closed forest cover is prevented by human activities (in particular periodic burning and heavy grazing), but together they cover extensive areas of Central and East Africa (including Madagascar), Central and South America, northern Australia and some of the drier parts of India and south-east Asia.

The climate of these areas is always warm, with little seasonal temperature variation. Amounts of annual precipitation are extremely variable and unpredictable and although in some areas the annual total may be fairly substantial, the rainfall is strongly seasonal in nature and all savannah zones experience a distinct dry season. The rains occur during the hottest season, when an

4.13   African tropical montane forest: (a) view of bamboo zone, Mount Kenya: (b) view of *Hagenia abyssinica* trees, tree-heathers and tussock grasses, Aberdare Mountains, Kenya.

average of 1000 mm may fall, and are followed by a long, dry season which may extend for more than half the year.

Savannah soils have a low humus content and, although they are not so heavily leached as those of the wetter tropics, laterisation is still the dominant pedogenic process. However, leaching is halted during the dry season and this, together with the return of minerals to the soil surface layers after grass fires during the dry season, helps to maintain soil fertility.

On the drier margins of tropical deciduous forest there is often a region of 'savannah woodland' in which, except in moist situations (i.e. depressions where the water table is nearer the surface), the trees are small and widely spaced. The herb layer of these woodlands and the more open, often treeless savannah grasslands is dominated by coarse, xeromorphic, fire-resistant grasses, members of the Cyperaceae (the sedge family), and some bulbous or tuberous perennials, which come into leaf and flower as soon as the rainy season commences. The grasses may form tall tussocks up to 4 m in height, but toward the drier margins of the savannah they are less abundant and lower-growing. Many savannah trees are deciduous and lose their leaves at the start of the dry season, but others, such as species of palm and pine retain their leaves. They all show marked xeromorphic adaptations, particularly to their leaves, which are reduced in size, and some species also possess a thick, corky, fire-resistant bark.

Savannah covers parts of the Pacific coast of Central America, the West Indies and is also characteristic of the 'llanos' of Venezuela and the 'campos' of Brazil, which lie to the south of the Amazon basin. The typical vegetation is a 'tall-grass' savannah, in which the grasses during the flowering season may exceed 1 m in height and include species of *Andropogon* (beard grass), *Aristida* (three awn), *Panicum* (millet), *Trachypogon*, *Sporobolus* and *Paspalum*. Small forbs belong mostly to the families Compositae and Fabaceae. The number of trees on the tall-grass savannah is variable and they are entirely absent from some areas, probably as a result of repeated burning. Some of the trees, such as *Curatella americana*, are evergreen and possess felty, hairy or waxy leaves as a protection against the arid environment. Species of palm of the genus *Copernicia* and *Pinus* spp. may also be present. In drier regions (i.e. those receiving less than 850 mm rain per annum) 'short-grass' savannah predominates, in which the clumps of xeromorphic grasses are less than 0.3 m in height; and, in the few limited regions of savannah where the rainfall

greatly exceeds 1000 mm annually, sedges (e.g. species of *Rhynchospora* (beak-sedge) and *Lagenocarpus*) become more abundant than the grasses.

In Africa, immediately surrounding the forest regions of the tropics, are areas of open savannah vegetation with widely spaced trees. Belts of this savannah run right across northern Africa between the scrub and semi-desert on the edges of the Sahara and the rain forests to the south, and around the forests of the Congo. Central Madagascar also has a cover of tropical grassland. As in South America, the African savannah has several distinct subdivisions dependent on the amount of annual precipitation received. In the most northerly areas approaching the borders of the Saharan desert where the rainfall is less than 500 mm per annum, the savannah grades into dry tropical scrub, but with increasing rainfall this is replaced by the so-called 'Sudan' savannah. Tall tussocks of *Pennisetum purpureum* (elephant grass), which may reach 4 m in height, are characteristic of these damper grasslands, as are several other fire-resistant grass species (e.g. *Panicum* (Millet), *Hyparrhenia* and *Andropogon* (beard grass)). In the Sudan savannah several species of *Acacia* trees predominate, along with large trees of *Adansonia digitata* (baobab), but with increasing rainfall these are replaced by larger, broad-leaved trees (principally species of *Isoberlinia*, *Lophira* and *Anogeissus*). To the south of the rain forests the savannah tree species are more numerous and include members of the Fabaceae, e.g. species of *Julbernardia* and *Brachystegia*.

4.7 Tropical scrub

Tropical xerophilous scrub, or 'thorn woodland', is intermediate between the savannah and desert formations in central Africa and Australia, where it is known as 'bush', and in northern South America, where the local name is 'caatinga' (white forest). It is also present locally in Central America and northern Africa. The climate of these regions is hot and dry, with an irregularly distributed rainfall of 400–900 mm per annum, and the aridity of the environment is further increased where the soil is coarse and fast-draining.

Tropical scrub vegetation is capable of withstanding long periods of drought. The leaves of many trees and shrubs are reduced to thorns or prickles and on other species the foliage is deciduous during the dry season. Some tree species also have specialised water-storage tissues, for example *Adansonia digitata* (baobab tree) and *Cavanillesia arborea* (Brazilian bottle-tree), which

4.14   East African savannah: (a) game range-land, Hell's Gate National Park, Kenya; (b) heavily grazed *Tarchonanthus camphoratus/Acacia* scrub near Lake Naivasha, Kenya.

store water in their large, swollen trunks. Succulent species are also a feature of tropical scrub and in very arid areas they may assume dominance.

The caatinga of South America is characteristic of north-east Brazil and northern Venezuela and Columbia. The dominant tree species are members of the Fabaceae and include *Mimosa* spp. and *Caesalpinia pulcherrima* (Barbados pride) in addition to the Brazilian bottle-tree and numerous succulents, including cacti, bromeliads and members of the Euphorbiaceae (the spurge family).

In Africa, thornbush, or 'bushveldt', covers vast tracts in East Africa, Somalia, Botswana, the Sudan and West Africa. The dominant trees are baobabs and smaller, thorny *Acacia* spp. which grow in association with shrubby succulents, including *Sansevieria* spp. (mother-in-law's tongue) and *Aloe* spp.

4.8 Desert

In desert regions less moisture is available for plant growth than in the dry grasslands on which they border and, as the grass species gradually disappear, they are replaced by desert scrub. The transition zone of semi-desert grades into true desert wherever the area of bare ground exceeds that covered by vegetation. Deserts are found in both the temperate and the tropical regions of the world and are of four types:

(a) Interior continental deserts, such as those of central and transcaspian Asia and parts of central Australia, which are beyond the reach of moisture-laden winds.

(b) Rain-shadow deserts, including the Great Basin desert in North America and the Patagonian desert in South America. These lie on the leeward side of high mountain ranges which intercept the rain-bearing winds.

(c) Subtropical deserts, e.g. the Saharan and Arabian deserts, which are situated within belts of high atmospheric pressure (the 'Horse' latitudes). Only the movement of fronts into these latitudes in summer may result in an occasional rain storm.

(d) Cool coastal deserts, such as the Atacama-Peruvian desert on the west coast of South America and the Kalahari-Namib desert on the south-west coast of Africa, which are almost without measurable rainfall and yet are regions of high relative humidity. Although these deserts lie within belts of high pressure, the proximity of cold ocean currents to the

land results in frequent fogs and mists, which favour epiphytic species and lichens.

Temperate deserts, as opposed to those of the tropics and subtropics, experience great seasonal as well as diurnal ranges of temperature. (Patagonia is an exception to this, since its proximity to the ocean largely prevents seasonal temperature extremes.) The winter months in these deserts are cold and, at high elevations, extremely so, with blizzards and frequent frosts. In parts of the Gobi desert, in central Asia, mean monthly temperatures below freezing may be experienced for half the year or more. During the summer months air temperatures are warm to hot – the average for the warmest months being between 21 and 30°C. Marked seasonal temperature variations are not a feature of tropical deserts, although large diurnal variations are, and nocturnal temperatures below freezing may be followed by a daytime maximum in excess of 50°C.

Average annual precipitation in desert regions is rarely in excess of, and usually well below, 250 mm, and occurs with such irregularity that many desert regions experience droughts of several years' duration. Low rainfall and high daytime temperatures result in very low relative humidities, which may be partly offset by fog and mist in coastal deserts, or, to a lesser degree, by the formation of nocturnal dew. Deserts are frequently subjected to strong winds, which may occur at any season and carry with them clouds of dust and sand.

In the most arid deserts there is no soil and the ground is covered with rocky debris or sand. With increasing rainfall, however, thin, coarse-textured sierozems (grey desert soils) may develop. These have a shallow surface horizon with a low organic content overlying a subsoil rich in accumulations of lime. They have very little moisture and, as a result of high evaporation, salts, particularly sulphates, may accumulate at the soil surface.

Many desert plants possess xeromorphic features which enable them to obtain, conserve and, if succulent, store water. A few very specialised desert plants are capable of surviving long rain-free periods by collecting the dew that condenses on them at night. After rain has fallen, small ephemeral plants appear; these germinate quickly from dormant, drought-resistant, buried seeds. Such species avoid the worst periods of drought and therefore tend to be less xeromorphic. Their life-cycle is short; flowering and fruiting is accomplished within two weeks in some cases. In a similar way, following rain, annuals emerge from underground tubers or bulbs and flower and fruit before the onset of another dry period. Desert ephemeral and annual species can remain dormant for several months, or even years, until the next rain storm.

The temperate deserts of Asia include those of Turkestan, which stretch eastwards over low land from the Caspian Sea to the mountains bordering China; the Dzungaria and Takla Makan deserts in the Sinkiang region of north-west China; and, one of the world's largest deserts, the Gobi and associated deserts, on the high plateaux of Mongolia and the Inner Mongolian region of China. To the north lie the 'desert steppes' – a transition zone between the grasslands and the desert, where the steppe grasses (*Festuca* spp. (fescues) and *Stipa* spp. (feather-grasses)) are gradually replaced by low bushes of *Artemisia* spp. (wormwood), *Atriplex canum* (salt bush) (on saline soils) and xerophytic desert grasses such as *Poa bulbosa* (bulbous meadow-grass).

In North Africa the tropical Saharan (the largest desert in the world) and Arabian deserts stretch from the Atlantic Ocean to the Red Sea and encompass vast expenses of sand. The vegetation is poorly developed and large areas, particularly those with mobile dunes, are entirely devoid of plant life. Long-lived desert plants include small *Acacia* trees and shrubs such as *Tamarix* spp. (tamarisk) and *Ziziphus* spp. There are only a few succulents present, but following rain a number of ephemeral species also appear (Fig. 4.15).

In western North America the temperate Great Basin desert occupies the entire state of Nevada and parts of the surrounding states to the east of the Sierra Nevada. It is a region of depressions and mountains. Xeromorphic shrubs are dominant over grass species and they include, as in Europe, species of *Artemisia* (wormwood) in addition to *Sarcobatus* spp. (greasewood) and *Atriplex*

spp. (salt bushes). *Populus* spp. (cottonwood trees) may be found in damper, low-lying areas. In the southern part of the Great Basin, *Cactus* spp. and communities dominated by *Larrea tridentata* (creosote bush) become increasingly common. These latter species are more typical of the hotter, subtropical deserts of North America, which extend from south-eastern California to Texas, and into Mexico. Here cacti, such as *Carnegiea gigantea* (saguaro), may attain the height of trees. Species of *Yucca*, including *Y. brevifolia* (Joshua tree), *Acacia greggii* (cat's claw) and *Fouquieria splendens* (ocotilla) are typical shrubs.

Temperate deserts in South America comprise the Monte-Patagonian deserts of Argentina and Bolivia, which occupy high-altitude, intermontane valleys in the Andes, and the Patagonian desert, which lies on high tableland to the east of the Andes and occupies most of southern Argentina. Woody, xerophytic, resinous scrub (the 'monte'), composed of members of the Zygophyllaceae in addition to coarse bunches of *Stipa* spp. (feather-grasses), covers much of the higher desert. Many species have reduced leaves or are leafless. In the Patagonian desert constant high winds, in addition to aridity, reduce the vegetation cover to widely spaced, low-growing clumps of grasses and cushion-forming shrubs. Patches of monte scrub occupy depressions and other sheltered positions.

On the western seaboard of South America lies the tropical Atacama-Peruvian desert, which has the lowest rainfall of any coastal desert in the world, although heavy fogs and high humidity are common features. This region is almost without any vegetation, but on seaward slopes, which receive some moisture from the mist and fog, sparse stands of the bromeliad *Tillandsia* (Spanish moss) and a few lichens occur.

More than half of Australia receives less than 250 mm rainfall per annum, and desert therefore makes up a sizeable proportion of that country, particularly in its central regions. The most characteristic plants are the very spiny *Triodia* spp. (spinifex grasses) and small shrubs of *Acacia* and *Eucalyptus* spp. On saline soils various halophytic members of the Chenopodiaceae (the goosefoot family), such as *Atriplex vesicaria* (salt bush) and *Bassia sedifolia* (blue bush), are present.

4.15   Sparse desert scrub vegetation in Oman.

4.9 Alpine vegetation

Alpine vegetation is found above the tree-line but below the permanent snow zone on high mountains throughout

4.16 (a) A high alpine scene in northern Italy – the Tre Cime de Lavaredo, 2999 m. (b) Tropical alpine vegetation on Mount Kenya – giant *Senecio* sp. and *Lobelia* sp. (foreground) growing at 5000 m.

the world. The altitude of the alpine zone varies with latitude and aspect (i.e. direction of slope); it is lower on north-facing slopes and at higher latitudes. The alpine climate is rigorous: at very high altitudes there is a reduction in precipitation and this, together with strong winds, facilitates rapid evaporation so that plants and soils dry quickly; there are large annual and diurnal temperature variations with, in particular, the possibility of frost at any time of the year, even during the growing season; intensity of solar radiation (especially of harmful ultraviolet rays) increases with altitude; and there is much winter snow and ice, some of which remains throughout the year on north-facing slopes. Alpine soils are thin and poorly developed. They include nutrient-deficient, acid rankers on hard, resistant rocks, and more fertile, thin rendzinas overlying limestone. On steep slopes and near the snow-line there is much bare rock and scree. The adaptations shown by alpine plants to their harsh environment and the montane vegetation of the British Isles are described in Chapter 9.

Alpine vegetation is found on the high mountains of the temperate regions of North and South America, Asia and New Zealand, but probably finds its best expression on high European mountain slopes. This vegetation is characterised throughout by a low-growing sward of grasses, sedges and abundant flowering plants, including many members of the Ranunculaceae, Rosaceae, Saxifragaceae and Gentianaceae. In situations where a deeper acidic soil has developed members of the Ericaceae, notably *Vaccinium* spp. (bilberry) and *Rhododendron* spp. are dominant; dense groves of *Rhododendron* spp. are particularly characteristic of the alpine zone of the central Asian mountains. On scree slopes and at the edge of the snow-line, lichens and mosses replace many of the vascular plants.

The lower limit of alpine vegetation on the highest mountains of tropical Africa, Asia and South America lies between 3000 and 4000 m above sea level and usually consists of a luxurious grassland with tall grasses, forbs and arborescent (tree-like) species. However, in drier situations and at higher levels the lack of water reduces the plant cover to a few sparsely scattered cushion- or rosette-forming species and tussocks of grass. Despite their equatorial location, the climate of these high mountains is cool with large diurnal temperature variation. Night temperatures usually drop below freezing, whilst daytime values may exceed 12–15°C.

In Central and South America the damp alpine meadows (the 'paramos') are dominated by tall tussock grasses and arborescent, pillar-like members of the Compositae, notably species of *Espeletia* and *Culcitium*, which may attain the height of small trees. On the drier slopes of the Andes and at high altitudes, the paramos border on the drier 'puna' vegetation, which is more xeromorphic in character. Here the low grasses grow together with cushion- and rosette-forming species and cacti. The plant cover is further reduced towards the permanent snow-line.

The lower zone of the African alpine belt (*c.* 3000–4000 m) supports ericaceous scrub dominated by large tree-heathers of the genera *Erica* and *Philippia* or plants with an ericoid habit, such as the composite *Stoebe kilimandscharica*. The bushes, which may attain 10 m in height, are often clothed with curtains of the epiphytic lichen *Usnea* spp. Towards the upper limit of this zone the heaths are reduced in size and scrub is replaced by tussock grassland, in which grow scattered tree-like *Dendrosenecio* spp. (giant groundsel) and *Lobelia* spp. (wooly candle-plant) up to 8 m in height. At higher elevations the arborescent species are replaced by lower-growing plants, including several *Helichrysum* spp. Over 4000 m moss-dominated communities, in which vascular plants are absent or scattered, are common. The thermal limit for plant growth is reached at about 5000 m.

Chapter 5

Woodlands, grasslands and heaths

Woodland is the natural, zonal climax vegetation of the British Isles, with the exception of exposed coasts and high mountain areas where severe climatic conditions, in particular strong winds, prevent tree growth. In these situations, woodland is replaced by low-growing grass or heath communities and, in very high rainfall districts of north-western Britain, by blanket bog (section 7.2.3). Elsewhere in Britain, however, grasslands and heathlands are semi-natural vegetation types which result from many centuries of human interference (grazing, mowing and burning), which has inhibited or prevented tree growth.

The present-day vegetation of the British Isles cannot be fully understood without reference to its development against the background of relatively recent geological and climatic changes. The starting point for the following historical account is the closing stages of the last glaciation of north-west Europe – the Weichselian – which commenced about 75 000 years before present (BP). As the ice sheets that covered most of the British Isles decayed, physical and chemical factors of the environment played the dominant roles in determining the composition, abundance and distribution of plant communities; but following the arrival of Neolithic agriculturists it became more and more difficult to separate abiotic effects from those induced by man.

5.1 Vegetation History

Palynological and stratigraphical analyses of post-glacial plant remains preserved in lake sediments and peats (section 7.1.7) have revealed the sequence of vegetational changes which commenced some 12 000 years ago at the end of the last ice age. As the climate ameliorated, various cold-tolerant plants migrated northwards into Britain and Ireland over land-bridges between southern Britain and continental Europe and western Britain and Ireland. The ice-free ground was colonised by tundra heath in which the low-growing or prostrate shrubs of juniper (*Juniperus communis*), dwarf birch (*Betula nana*) and dwarf willow (*Salix herbacea*) were joined by a variety of cold-tolerant herbs, grasses and sedges. Some of these plants may also have survived the last glaciation in ice-free areas in southern Britain (section 9.3).

By approximately 10 000 BP climatic improvement had encouraged open birch forest (mainly of *Betula pubescens*, but also containing some *B. pendula*) to spread north over the tundra heaths, whilst in southern and eastern England pine (*Pinus sylvestris*) had established. Additional trees and shrubs included hazel (*Corylus avellana*), aspen (*Populus tremula*), rowan (*Sorbus aucuparia*) and willows (*Salix* spp.). This Pre-Boreal period lasted for approximately a thousand years

and was succeeded by the warmer Boreal period (Fig. 5.1), during which forest expanded to cover most of the British Isles. Pine extended northwards into the birch woodlands, reaching the Scottish Highlands by the close of the Boreal period, whilst hazel also expanded rapidly, probably forming an understorey beneath the pine. At some point approximately half way through the Boreal (*c.* 9000 BP), the climate improved further and broad-leaved forest trees – oaks (*Quercus petraea* and *Q. robur*), elms (*Ulmus* spp., predominantly *U. glabra*) and alder (*Alnus glutinosa*) – made their appearance in southern and western Britain. All of these species probably reached Ireland before the land-bridge (or bridges) was sundered by rising sea levels.

By the end of the Boreal period, the lowlands of Britain and Ireland were covered with a closed canopy of broad-leaved woodland, whilst the north was still forested with birch and pine. The transition between

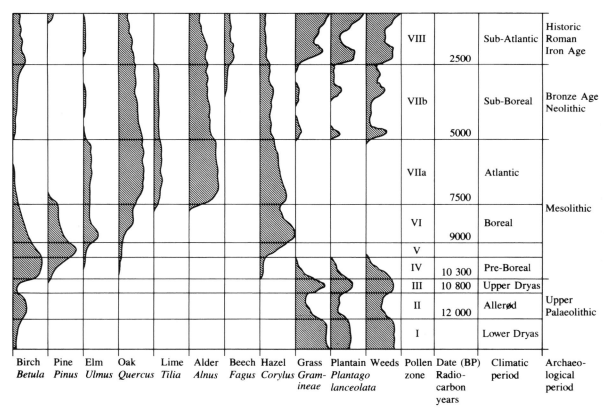

5.1　Generalised pollen diagram for the late-Devensian and Flandrian periods in England, showing the pollen curves for selected pollen types expressed as percentage arboreal pollen. Zones are according to Godwin (1975) and superimposed are radiocarbon dates, climatic periods and archaeological cultures (from Moore & Webb, 1978).

the Boreal and Atlantic periods (*c.* 7000 BP) was marked by a considerable rise in both temperature and rainfall; the mean summer temperature was perhaps 2.5°C higher than it is now and the growing season was longer. Oak, elm, alder, lime (mainly *Tilia cordata*, the small-leaved lime), holly (*Ilex aquifolium*) and to a lesser extent ash (*Fraxinus excelsior*) all expanded during this period, whilst pine and birch declined and became infrequent outside central and northern Scotland. About 6000 BP rising sea levels submerged the remaining land-bridge between Britain and the continent of Europe; Britain was now an island.

The increased rainfall of the Atlantic period promoted the development of ombrotrophic raised mires in lowland lake basins and on coastal flats, whilst on the uplands of northern and western Britain, blanket mire development was initiated, replacing the former pine and birch forests. The fossil record for this period shows a dramatic increase in the remains of peat-forming plants – *Sphagnum* spp., *Eriophorum* spp. and members of the Ericaceae were all abundant wherever peat formation had been initiated (section 7.1.5).

By the end of the Atlantic period (*c.* 5000 BP) almost the whole of Britain was covered with virgin forest or 'wildwood' (Rackham, 1976). Large mammals inhabited the forest (deer, wolves and wild oxen) as did small nomadic tribes of Mesolithic hunter-gatherers. But the wildwood was, as yet, untouched to any significant extent by man. Some of the regional variation in the wildwood of the Atlantic is still evident in the composition of present-day woodlands. Pine, for example, was restricted to central and northern Scotland, whilst further south and along the western seaboard oak was the principal forest tree. Mixed deciduous woodland dominated by oak, lime and elm with abundant hazel and alder and a smaller amount of ash covered the whole of England and Wales – probably reaching the lower hill summits. Ireland also had a similar forest cover, but lime and ash were largely absent and elm and hazel more abundant. Outside the central plain, however, pine appears to have persisted. (Birks *et al.*, 1975).

The end of the Atlantic period is marked throughout most of Europe by a sudden and synchronous drop in the quantity of elm pollen. This 'elm decline' was originally attributed to a combination of climatic change and the arrival of Neolithic agriculturalists, who may have selectively favoured elm for cattle fodder (Iversen, 1949; Troels-Smith, 1964). However, more recent evidence points to an early attack of a fungal wilt disease

similar to the Dutch elm disease which has led to the loss of so many elm trees at the present time. Whatever the causal factor or factors, this elm decline in pollen diagrams marks the transition from the wet, warm climate of the Atlantic period to the cooler, drier, more continental climatic conditions of the Sub-Boreal which followed. It also marks the introduction of settled agriculture to Britain and the onset of large-scale interference with the woodland cover by Neolithic man. Areas of forest were cleared on the lighter sandy or chalky soils of the uplands, probably by a combination of felling and burning, to provide land for cultivation and grazing. Palynological and stratigraphical analyses of sediment profiles for this period often reveal charcoal layers concomitant with a sudden drop in the tree pollen frequencies (although pioneer species such as birch and alder appear to have increased as a result of woodland disturbance), whilst non-tree pollen values, including herbs and grasses, increase. Even more conclusive evidence of primitive agriculture is provided by the appearance of cereal pollen (notably wheat and barley) and a substantial increase in the pollen curves of plants associated with open habitats, including plantains (*Plantago* spp.), wormwoods (*Artemisia* spp.) and docks (*Rumex* spp.).

Although natural, broad-leaved, deciduous woodland remained widespread throughout the Sub-Boreal, changes in the species composition occurred. In the mixed forest elm was now infrequent, but oak, lime and hazel maintained their former levels. Ash, however, increased throughout most of the British Isles, in particular in southern Scotland and northern England, whilst beech, and to a lesser extent hornbeam, both of which probably entered Britain during the late Atlantic, expanded in southern England. Neolithic deforestation was also responsible for the appearance, for the first time since the Pre-Boreal, of lowland heaths, which may have formed on soils of low base status following temporary cultivation. The loss of soil fertility from woodland clearance and crop production was perhaps equivalent to a much longer period of natural leaching of the soils by rainwater, and podzolisation could have preceded the eventual spread of ericaceous shrubs, in particular *Calluna vulgaris* (heather), (Dimbleby, 1962). Some chalk grasslands also have their origin in the Sub-Boreal. Pollen diagrams for Kent, for example, suggest that the chalk hills of south-eastern England had been largely cleared of forest by the end of the Neolithic (*c.* 3700 BP) and instead supported herb-rich grassland

(Godwin, 1962). Grazing may also have been responsible for an increase in the abundance of spiny, unpalatable shrubs at this time, e.g. *Ulex* spp. (gorse). Thus, although the impact of Neolithic farming must have varied from place to place, there was sufficient activity in some localities to create and maintain areas of grass and heath.

The Sub-Boreal was succeeded in 2800 BP by the moister Sub-Atlantic, which continues to the present day. The start of the Sub-Atlantic coincides with the establishment of Celtic Iron Age culture in Britain and from the onset the vegetation was increasingly under man's influence and control. Although the proportions of most forest trees show little alteration in the early Sub-Atlantic, birch, ash, beech and hornbeam all expanded. The increase in the latter two species may have been a response to the warmer climate of this period. However, it is equally plausible that all four were pioneer species of secondary woodland on abandoned agricultural land. One response to the wetter conditions of the Sub-Atlantic was the resurgence in mire growth, succeeding a general decline during the drier Sub-Boreal. Peat profiles of this period reveal a layer of wet *Sphagnum*-rich peat overlying the drier Sub-Boreal horizons in which heathers are more abundant. Elsewhere, in deforested areas with sandy, fast-draining soils the higher rainfall may have encouraged podzolisation and the extension of *Calluna*-dominated heathland.

During the early Iron Age the rate of deforestation increased, with the aid of improved metal tools; timber was required for housing, heating, implement manufacture and for charcoal to fuel primitive iron smelters. Land previously cleared of trees was kept open by cultivation or pastoral activities and sheep-grazing increased, particularly in the north and west of Britain.

When the Romans entered Britain in AD 43 the resident population probably stood at less than half a million (Hoskins, 1955) and the main centres of population were still on the lighter upland soils. The Romans intensified agriculture and, by road building, opened up previously isolated areas. Pollen analyses indicate that extensive areas of southern England were arable at this time, whilst much of northern and western Britain was still wooded. Extensive drainage works in the Fens permitted an expansion of cereal cultivation in eastern England, whilst further woodland clearance accompanied the growth of iron smelting in the Weald of Kent and the Forest of Dean. The Romans were also responsible for introducing many new crop plants to

Britain, including several trees: walnut (*Juglans regia*), plum (*Prunus domestica*), crab apple (*Malus sylvestris*) and sweet chestnut (*Castanea sativa*).

By the end of the Roman occupation, however, the forests on the heavy clay lands of the Midland plain, parts of East Anglia and the Weald (outside the clearings around the smelters) must still have been largely untouched. The cultivation of these areas required larger, more substantial ploughs, whose use became widespread in Anglo-Saxon times (AD 450–1066). Some of the agricultural land on the uplands was abandoned and returned to scrub and woodland, whilst the Fens were inundated by a marine transgression and former farmland was once more swamp. Godwin (1975) suggests that the expansion of beech on the chalk hills of southern England at this time was probably due to a shift in arable cultivation on to the heavier and deeper soils of the lowlands, leaving the thin chalk soils which had previously been cleared open to swift colonisation by beech.

During the 600 years of Anglo-Saxon settlement, England became a land of villages and fields (Hoskins, 1955). Around each village land was cleared of woodland to form two or three large open fields divided into strips. Ploughing of these strips produced high ridges separated by furrows; evidence of former ridge-and-furrow ploughing is still visible under permanent pasture in the Midlands landscape today.

The Norman invasion of 1066 was followed by William the Conqueror's great Domesday Survey 22 years later. Although the survey was not entirely comprehensive (there is no information on Northumberland or Durham and very little on Cumberland and Westmorland), it appears that approximately 35% of England at that time was arable, 15% woodland and wood pasture (woodland in which domestic animals grazed), 30% pasture (grazing land) and 1% meadow (grassland cut for hay). The remaining 20% was made up of houses, gardens, moor, heath and mountain (Rackham, 1986). Thus, by the turn of the twelfth century, the former 'wildwood' was much reduced and most remaining woodland was managed for timber and wood (see below). Large tracts of land did, however, become 'forest' in a legal sense at this time: areas were set aside as royal game preserves for the Norman kings and became subject to forest law. The forest boundaries may have encompassed woodland, although they were just as likely to contain unwooded moor (e.g. Pickering Forest), heath (e.g. Sherwood Forest) or former arable land (e.g. the New Forest). Although, by law, the King's permission was required for tree felling or woodland clearance within a royal forest, in practice this was rarely done. Further impoverishment of the woodland cover resulted from overgrazing by domestic animals and the failure of damaged tree seedlings to regenerate. The Normans were also responsible for the introduction of the rabbit (*Oryctolagus cuniculus*) into Britain as a source of meat and fur. Although originally confined to 'warrens', in later centuries they spread and became an important biotic factor by grazing and preventing tree regeneration. (Sheail, 1971).

This period also saw the rise of the monasteries, which controlled large areas of land. The Cistercians, in particular, founded houses in remote locations (e.g. Fountains and Rievaulx Abbeys in Yorkshire and Furness Abbey in Cumbria). The monks were efficient farmers – especially of sheep – and during the twelfth and thirteenth centuries a massive expansion of pastoral farming took place on the chalk and limestone uplands of Yorkshire and Lincolnshire, the Lake District fells and the moorlands of Wales and southern Scotland. Pollen records for some of these areas provide evidence of a rapid increase in grassland at this time with a consequent decrease in woodland, reflecting the destructive effect that prolonged sheep grazing has on woodland regeneration.

From the fourteenth and fifteenth centuries onwards the fenlands of eastern England underwent renewed drainage and became a prosperous agricultural region, supporting large flocks of sheep and herds of cattle. The Tudor period also saw a substantial increase in the demand for woodland products, in particular oak timbers for the English navy. Soon there were complaints of a timber shortage, which heightened during Stuart times and culminated in the Admiralty consulting the newly formed Royal Society. The result was the appearance of John Evelyn's *Sylva: a Discourse of Forest Trees, and the Propagation of Timber* in 1664, which advocated a policy of replanting as a means of making good previous losses. (Darby, 1951).

By 1700 little remained of the natural British vegetation; woodland was replaced by arable or pasture land and large areas of wetland had been drained (Ratcliffe, 1984). The open field system of agriculture which operated over much of central England declined during the eighteenth century and was replaced by a patchwork of smaller fields enclosed by hedges. The Parliamentary Enclosure Acts of 1750 onwards hastened this change. Besides the open arable fields, ex-

tensive 'wastes' were enclosed; millions of acres of heath and moorland were brought under cultivation. Following enclosure and fertilisation with lime, vast areas of heath in Norfolk and Lincolnshire were converted to cereal farming. (Hoskins, 1955).

Evelyn's *Sylva* initiated some tree planting, but it was not until the late eighteenth century onwards that plantation forestry really expanded – existing woodlands were extended and new plantations created. Oak, ash, beech and pine were planted (the last two often in areas outside their natural range), plus sycamore (*Acer pseudoplatanus*), which had been introduced from central Europe in the late sixteenth or early seventeenth century. Other trees introduced at about the same time were the European larch (*Larix decidua*), Norway spruce (*Picea abies*) and, from southern and central Europe, silver fir (*Abies alba*). Their use as forestry trees increased greatly during the nineteenth and twentieth centuries. Afforestation of land with poor soils of low agricultural value (e.g. lowland heath and upland grassland) was accelerated following the creation of the Forestry Commission in 1919. The start of wide-scale afforestation in Britain is marked at the top of many pollen diagrams by an appreciable rise in the pollen values for beech, pine and non-native conifers.

Up until the twentieth century agriculture was managed on a system of 'low input-low output', which by modern standards would be considered inefficient, and many farmland plant communities were floristically diverse (Ratcliffe, 1984). Flower-rich flood meadows, for example, were fertilised naturally by nutrient-rich alluvium deposited by rivers and streams which overtopped their banks during the winter months and flooded the low-lying land. During late spring and summer the waters subsided and the lush growth of meadow plants was removed as a crop of hay. After the Second World War agricultural practices changed dramatically and became increasingly intensive; ancient pasture, lowland heath and chalk grassland were ploughed; low-productivity semi-natural grass swards were resown with more productive seed mixes, following the application of selective herbicides and inorganic fertilisers; woodlands were grubbed out and converted to arable land or plantation; wetlands, including many flood meadows, were reclaimed. The growth of urban and industrial areas also resulted in the loss of more natural and semi-natural vegetation. This rate of loss has accelerated dramatically during the last 50 years. For example in 23 counties of England and Wales 45% of ancient, semi-natural woodland was lost during the period 1933–83, either through grubbing out to provide more farm-land or conversion to conifer plantation. Only 76 500 hectares remain. During the period 1950–84 40% of lowland heaths were lost, largely by conversion to improved grassland or arable, afforestation or building. Lowland chalk and limestone grasslands have suffered an 80% loss since 1940, again largely by conversion to arable or improved grassland, whilst only 3% of lowland neutral grasslands, including herb-rich hay meadows, have been left unaltered by agricultural intensification. (Nature Conservancy Council, 1984).

5.2   Woodlands

The diversity of woodland types in the British Isles at the present time still largely reflects local variations in soil and climate. In general, however, owing to the greater degree of destruction and exploitation in the past, British woods are much more fragmentary and less well defined in floristic terms than are the woods of the remainder of Europe. As a result, it is often difficult to identify typical British examples of each European association. The phytosociology of British woods, although partly covered by several European authors (Braun-Blanquet & Tüxen, 1952; Klötzli, 1970), has not been studied systematically and much work still requires to be carried out. The most informative recent work on British woodland types is that of Peterken (1981).

Relics of primeval woodland which have continuously occupied the same site since the time of the Atlantic wildwood are referred to as 'primary woodland'. Although the woods may have been managed for centuries, they retain trees and shrubs native to the sites, which have never been clear-felled or replanted. 'Secondary' woodlands have formed on land which was once cleared of trees, although deforestation may have taken place tens or even hundreds of years ago. Eaton Wood in Nottinghamshire is an example of an old, secondary woodland which colonised abandoned arable land some time before the Norman Conquest; evidence of ridge-and-furrow agriculture is still visible beneath the trees.

On the basis of its age, a woodland may also be defined as 'ancient' or 'recent'. All ancient woodlands date back to medieval times or earlier (i.e. pre AD 1600); recent woodlands are post-medieval. Secondary woodlands may be ancient or recent, whilst all primary woods are ancient.

Ancient woodlands generally have a richer flora than recent woods and certain plants appear to be confined to the older sites. In a survey of Lincolnshire woodlands Peterken (1981) included the following in a list of 'indicator plants' of ancient woodland: *Anemone nemorosa* (wood anemone), *Lamiastrum galeobdolon* (yellow archangel), *Oxalis acetosella* (wood sorrel), *Paris quadrifolia* (herb Paris), *Sorbus torminalis* (wild service tree) and *Tilia cordata* (small-leaved lime).

5.2.1   *Woodland management*

As a result of the long history of man's interference with the British vegetation, the woodlands have all been subject to human modification and disturbance and most have been managed for centuries to supply woodland products. Formerly, the most widespread management scheme was the coppice system, which relies on the inherent ability of nearly all native British trees to grow again from a cut stump. This stump (the stool) produces shoots (poles) which, when large enough, may be cut to yield a crop of small wood (coppice) for use as fuel, fencing or other light construction work. Trees for larger timber (planks, etc., for heavy construction) were not coppiced. These standing trees, or 'standards', were often grown amongst the coppiced 'underwood' producing a woodland of coppice-with-standards (Fig. 5.2).

The coppice system may have been employed by Neolithic peoples. Evidence for this early woodland management is provided by various wooden trackways preserved in the peat of the Somerset Levels. Tracks constructed of both underwood and timber suggest that coppicing was already practised at that time. (Coles & Hibbert, 1968; Coles *et al.*, 1970). From medieval times onwards most woodlands were of the coppice-with-standards type; the underwood was coppiced on a rotation of usually five to eight years (less frequently coppice cycles of 20 or more years are recorded), providing successive crops of wood, whilst standards were felled less frequently, as and when timber was required. These managed woodlands formed an integral part of the local village economy. In the underwood various native woodland trees and shrubs were coppiced: oak, ash, hazel, maple, elm and lime, whilst oak appears to have been the commonest timber tree.

In essence, the coppice system of woodland management persisted well into the late nineteenth and early twentieth centuries, although coppice 'improvement' by planting of both native and non-native trees (e.g. sweet

a

b

5.2 Coppiced woodland in Nottinghamshire: (a) newly coppiced area in Eaton Wood showing cleared understorey. Standards are ash, birch and oak; coppice is ash and hazel. (b) Part of Treswell Wood three years into the coppice cycle. Note the dense growth of *Rubus fruticosus* and *Deschampsia cespitosa* in the field layer.

chestnut) became increasingly common from the seventeenth century onwards (Peterken, 1981). The main decline of coppice management did not start until the late nineteenth century when, for largely economic reasons, this form of woodland management fell from favour. Many woods were replanted with conifers, whilst others were retained as cover for game and foxes. Today only a few woods are managed by traditional methods, yet most remaining semi-natural woodlands owe their existence to their former value as sources of renewable fuel and timber for the local community.

5.2.2 *Woodlands on acid soils*

The dominant tree of British woods on acid soils is either *Quercus petraea* (sessile oak), *Q. robur* (pedunculate, -oak). *Pinus sylvestris* var. *scotica* (Scots pine), *Fagus sylvatica* (beech), *Betula pendula* (silver birch) or *B. pubescens* (downy birch) or a combination of some of these. The principal classes of woodland vegetation in this category are Quercetea-robori-petraeae and Vaccinio-Piceetea. (Alder swamp woodland and willow carr, which develop under the influence of a high water table on substrates rich in organic matter, are dealt with in sections 6.5.7 and 6.5.8.)

5.2.2.1   *Deciduous woodlands on acidic soils (Quercetea-robori-petraeae)*

This class, and the single order Quercetalia-robori-petraea and alliance Quercion-robori-petraeae, comprise all oakwoods on acid soils which occur throughout Britain from Devon in south-west England to Wester Ross in north-west Scotland. They form the characteristic woodland vegetation of west Wales, the Lake District, the Pennines and south-west, central and much of northern Scotland, but are largely absent from the heavy clays of eastern and central England. In common with several other accounts of woodland phytosociology, the less widely distributed beechwoods on non-calcareous soils in southern Britain are also included within this group.

Oak-birch woodlands

Quercus petraea is indicative of soils of more acid reaction than *Q. robur* and, apart from a few lowland areas, the former is a tree of upland or northern oakwoods on shallow, siliceous soils under extreme conditions of rainfall, slope, aspect and topography. *Quercus robur* is more characteristic of lowland, acid woodlands on podzolised sands and gravels, where occasionally, in the south of England, it has been replaced by *Fagus sylvatica*. However, *Q. robur* grows at a height of 400 m in Wistman's Wood, Devon. Where both species of oak occur together, as in Sherwood Forest, Nottinghamshire, hybrids between them are frequent. Many oak-birch woods have been managed as coppice oak or hazel, but in most of them management has now ceased. Upland woodlands are often grazed by sheep; this limits both tree regeneration and the full development of the field layer.

Alongside the *Quercus* spp. the only other trees which form a major part of the woodland canopy are *Betula pendula* and *B. pubescens*, which often replace oak after felling; *Sorbus aucuparia* (rowan) and *Ilex aquifolium* (holly) are frequent. Most woodlands, especially those on nutrient-poor soils, support relatively few shrub species. *Corylus avellana* (hazel) grows in woods with deeper, nutrient-rich soils; and *Salix* spp. (willows) are often present in wet areas or beside streams. The woody climbers *Hedera helix* (ivy) and *Lonicera periclymenum* (honeysuckle) are common.

Several variants of the herb layer flora can be

5.3   Woodland on acidic soils: sessile oakwood (Quercetea-robori-petraeae) at Coed Cymerau, north-west Wales. Note the luxuriant ground cover of bryophytes and ferns.

identified. On well-drained soils in upland areas or on podzolised sandy soils in lowland Britain the heath-like ground vegetation is dominated by *Vaccinium myrtillus* (bilberry), *V. vitis-idaea* (cowberry), *Calluna vulgaris* (heather), *Anthoxanthum odoratum* (sweet vernal-grass), *Deschampsia flexuosa* (wavy hair-grass) and *Holcus mollis* (creeping soft-grass). Less frequent associates are *Melampyrum pratense* (common cow-wheat), *Galium saxatile* (heath bedstraw), *Potentilla erecta* (common tormentil) and *Blechnum spicant* (hard fern). Clearings or felled areas quickly become colonised by *Rubus fruticosus* (bramble), especially in lowland woods, or *Pteridium aquilinum* (bracken), both of which reduce the light intensity reaching the woodland floor. Where either of these two species occurs in abundance they exclude most other plants. Occasionally, the delicate *Corydalis claviculata* (climbing corydalis) climbs up the *Pteridium* fronds to reach the light.

On more fertile soils, *Corylus avellana* occurs in the understorey and the ground flora is more diverse. Additional plants may include *Anemone nemorosa* (wood anemone), *Hyacinthoides non-scripta* (bluebell), *Lysimachia nemorum* (yellow pimpernel), *Veronica chamaedrys* (germander speedwell), *Primula vulgaris* (primrose), *Oxalis acetosella* (wood sorrel) and *Viola riviniana* (common dog violet). *Deschampsia cespitosa* (tufted hair-grass) and *Dryopteris* spp. (buckler ferns) occur in damper areas.

In many upland woods streams and flushes enrich the soil, which thereby supports plants more typical of lowland damp woods and wetlands, for example *Caltha palustris* (marsh marigold), *Ajuga reptans* (bugle), *Geum rivale* (water avens), *Juncus articulatus* (jointed rush) and *Ranunculus repens* (creeping buttercup).

There is considerable bryophyte diversity in these oak-birch woodlands, which increases with rainfall. The principal ground-inhabiting bryophytes are the mosses *Rhytidiadelphus loreus*, *Dicranum majus*, *Plagiothecium undulatum*, *Polytrichum formosum*, *Pleurozium schreberi* and the liverwort *Bazzania trilobata*. On rocks and fallen tree trunks *Isothecium myosuroides* and *Dicranum scoparium* predominate. Tree bases support abundant *Mnium hornum* and occasionally *Dicranum fuscescens*, whilst on the trunks *Isothecium myosuroides* and *Hypnum mammillatum* occur in abundance. The epiphytic bryophytes of the higher parts of trunks and branches include species of *Grimmia*, *Pottia*, *Orthotrichum* and *Tortula*. Streamside rocks and river banks also support a very rich bryophyte flora, the most obvious and abundant species of which are *Polytrichum commune*, *Thamnobryum alopecurum*, *Hyocomium armoricum*, *Brachythecium rivulare*, the thallose

a   b

5.4 Woodland species: (a) *Corylus avellana*, a common understorey shrub of oak and birch woods on richer soils; (b) *Anemone nemorosa*, an early spring-flowering woodland herb.

liverworts *Pellia epiphylla*, *Marchantia polymorpha* and several species of leafy liverwort. Where drainage or run-off of water is impeded the bog mosses *Sphagnum palustre*, *S. subnitens* and *S. recurvum* often occur in quantity.

Beechwoods on sands and podzols

Acid beechwoods occur on non-calcareous soils: the sands and gravels of the Chiltern Plateau; sands of the Lower Greensand series on the fringes of the London and Hampshire basins; the Bagshot and related sands overlying the London clay in the same basins, and their associated gravels. These woods are mainly small in extent. In some woods there are pure stands of *Fagus sylvatica* (beech), whilst in others there is a mixture of beech with *Quercus petraea*, *Q. robur*, *Betula pubescens* and *B. pendula*, the latter two occurring in openings in the beech canopy. *Ilex aquifolium* and *Sorbus aucuparia* may also be common.

Owing to the dense tree canopy very few accessory species grow beneath the beech, and the acid soils with their accumulated litter give rise to leached podzols supporting the more acid-loving species *Vaccinium myrtillus*, *Pteridium aquilinum*, *Anthoxanthum odoratum*, *Holcus mollis* and *Deschampsia flexuosa*, plus *Rubus fruticosus* and the ferns *Blechnum spicant* and *Dryopteris* spp. Small herbs include *Anemone nemorosa* and *Oxalis acetosella*. The ground layer is patchy and contains the cushion-forming mosses *Polytrichum formosum*, *Dicranella heteromalla*, *D. scoparium*, *Mnium hornum* and *Leucobryum glaucum*, together with the leafy liverwort *Diplophyllum albicans*.

5.2.2.2 Northern pine and birch woodlands (Vaccinio-Piceetea)

The single order of this class, the Vaccinio-Piceetalia, includes two alliances of woodland vegetation on acid, podzolic soils: Vaccinio-Piceion, containing associations of pine woodlands, and Betulion pubescentis, the northern birchwoods.

Pinewoods (Vaccinio-Piceion)

Pinus sylvestris is the dominant forest climax tree of the central and northern Highlands of Scotland on acid, podzolised soils with mor humus. It replaces oak-birch woodland away from the coast and on mountain slopes. Formerly, *Pinus sylvestris* dominated the landscape of Scotland north of the valleys of the rivers Forth and Clyde as far north as Wester Ross. Nowadays, however,

it is restricted to small stands which are only remnants of its former extent. Man's activities have reduced the Old Caledonian Pine Forest of antiquity to the point of extinction. The southern part of the present distribution is seen as only a few isolated trees in Glen Falloch, north of Loch Lomond. Better preserved examples occur further north at Tyndrum, Loch Tulla and Rannoch Moor in Perthshire, Rothiemurchus in Inverness-shire, Loch Maree in Ross-shire and Ballochbuie on Deeside.

The pinewoods of the western Highlands are situated in an area of high rainfall and their ground vegetation (especially the bryophytes) resembles that of the sessile oakwoods of the same region. To the east, however, on drier soils the under-tree vegetation is similar to dry heath (section 5.4). The tree and shrub layer is dominated by *Pinus sylvestris* which is occasionally associated with *Betula pubescens*, *B. pendula*, *Sorbus aucuparia*, and, in the west, *Ilex aquifolium*. *Juniperus communis* (juniper) is common and often forms a dense under-tree layer in the drier eastern pinewoods.

In the wetter western pinewoods the principal field layer species are *Vaccinium myrtillus*, *Molinia caerulea* (purple moor-grass), *Deschampsia flexuosa* and *Blechnum spicant*. In the east, *Calluna vulgaris* is co-dominant with *Vaccinium myrtillus* and *V. vitis-idaea*. An additional and frequent component of the field layer is *Pteridium aquilinum*.

Several uncommon flowering plants are characteristic of pine forest, including *Moneses uniflora* (one-flowered wintergreen) and the orchid *Goodyera repens* (creeping lady's tresses). The bryophyte layer is well developed, particularly in the west, where it is frequently dominated by species of *Sphagnum* and a large number of associated mosses and liverworts, most of which also occur in the *Quercus petraea* woodlands. *Ptilium crista-castrensis*, a feather-moss of localised distribution, occurs in these woodlands.

Birchwoods (Betulion pubescentis)

Native birchwoods occur in the north of Scotland, where they replace Scots pine forest in Caithness and Sutherland, and on mountain sides further south above the altitudinal limit of pine. Occasionally *Corylus avellana* replaces birch in the extreme north of Scotland. Further south in Britain, birch woodland also occurs as secondary woodland following disturbance.

A greater variety of trees occurs in birch woodland than in the related pinewoods, and apart from *Betula pubescens* and *B. pendula* these commonly include *Sorbus aucuparia* and *Corylus avellana*. *Populus tremula*

5.5 Scots pine forest (Vaccinio-Piceion): remnant of Old Caledonian Pine Forest beside Loch Clair, Glen Torridon, Scotland.

(aspen), *Prunus padus* (bird cherry) and *Crataegus monogyna* (hawthorn) are of more local occurrence, whilst in wet places several *Salix* spp. (willows) form pure stands. As in the related pinewoods, *Juniperus communis* is a frequent component of the understorey in north-eastern Scotland.

The field layer is very similar to that of the pinewoods, except that there is a greater abundance of grasses compared with the ericaceous shrubs of the former. This is probably a result of edaphic differences, and in particular the moder humus under the deciduous trees compared with the very acid mor humus of the pine litter. The most abundant field layer plants are the grasses *Agrostis capillaris* (common bent), *Anthoxanthum odoratum, Deschampsia flexuosa, Festuca ovina* (sheep's fescue) and *Holcus mollis*, together with the small herbs *Anemone nemorosa, Galium saxatile, Luzula pilosa* (spring woodrush), *Lysimachia nemorum, Oxalis acetosella, Potentilla erecta*, and *Viola riviniana*. Less frequent plants on peaty soils are the small orchid *Listera cordata* (lesser twayblade) and *Trientalis europaea* (chickweed wintergreen), both of which may also be found in pinewoods.

5.2.3 Lowland woodlands on nutrient-rich soils (Querco-Fagetea)

Woodlands of lowland Britain on brown earths or calcareous brown earths over clay, marl, calcareous sand, chalk or limestone belong to the class Querco-Fagetea, which embraces all of the mixed deciduous British woodlands of tall forest trees. This class is represented extensively throughout Europe and alliances include:

(i) Alno-padion – alderwoods of western and central Europe on damp, humus-rich, mineral soils.

(ii) Fagion sylvaticae – beechwoods. This is the dominant forest type within the temperate forest formation of central Europe on moist, calcium-rich soils.

(iii) Ulmion carpinifoliae – mixed oak-elm valley woods on fertile soils from western Europe eastwards to Poland. Although woodlands of this type may once have been present along lowland river valleys in Britain, the land has long since been cleared of trees and only very small areas remain. These woodlands are so fragmentary and poorly defined that they are not dealt with in further detail.

(iv) Fraxino-Brachypodion – a provisional alliance containing woodland associations with a high *Fraxinus excelsior* component in the canopy, on the limestones of central and northern Britain.

(v) Carpinion betuli – mixed, deciduous oak-hornbeam forests on fertile brown earths. This alliance is Eurasian in distribution with a wide geographical range covering very large areas of central and eastern Europe to the Balkans. In Britain hornbeam has a limited distribution and is absent from many woodlands in this category.

a

b

5.6   (a) *Juniperus communis*, a common under-shrub of the pinewoods of north-eastern Scotland. (b) *Trientalis europaea*, a characteristic ground floor herb of northern birchwoods and Caledonian Pine Forest, growing here amongst *Vaccinium myrtillus*.

5.2.3.1   *Alderwoods on mineral soils (Alno-padion)*

Alderwoods on fertile soils with a weakly acid, neutral or alkaline reaction may form the edge communities of other woodland types in areas which are damp or subject to occasional waterlogging. These woods are of very localised occurrence in Britain and are rarely extensive; they differ from the alderwoods of the class Alnetea glutinosae (section 6.5.7) which form on waterlogged organic substrates.

The tree layer consists of a number of species in addition to *Alnus glutinosa*, including *Fraxinus excelsior, Betula pubescens* and *Prunus padus*. The commoner shrubs are *Corylus avellana, Crataegus monogyna, Sambucus nigra* (elder), *Salix cinerea* (grey willow) and *Viburnum opulus* (guelder rose). The woody climber *Lonicera periclymenum* may also be present.

The field layer is frequently dominated by *Rubus fruticosus, Pteridium aquilinum* or *Dryopteris* spp., although *Ajuga reptans, Mercurialis perennis, Ranunculus repens* and *Filipendula ulmaria* (meadow-sweet) are usually present. Additional species which may be found are *Allium ursinum* (ramsons) on alkaline soils; the grasses *Agrostis* spp., *Deschampsia cespitosa* and *Holcus mollis*; and sedges, notably *Carex pendula* (pendulous sedge), *C. remota* (remote sedge) and *C. sylvatica* (wood sedge).

These damp alderwoods provide a suitable, moist habitat for the growth of bryophytes on the soil surface and on tree bases, trunks and branches. Ground-inhabiting species include *Brachythecium rutabulum, Bryum capillare, Atrichum undulatum* and *Eurhynchium praelongum*. Those commonly found on tree bases are *Mnium hornum, Plagiothecium denticulatum* and *Hypnum cupressiforme*, whilst *Orthotrichum affine* and *Zygodon viridissimus* are epiphytic.

5.2.3.2   *Beechwoods on chalk and limestone (Fagion sylvaticae)*

Beech woodland may originally have spread following deforestation and eventual neglect of agricultural land on the light chalk soils of southern England during the early part of the last millennium. Beechwoods of the Fagion sylvaticae now occur on the chalk and limestone rendzina soils of southern England and south Wales; on fertile brown earths, beech and oak (*Quercus robur*) grow together. The relative abundance of these two trees in the canopy is determined partly by soil preference and partly by management regime: beech does not grow well on heavy clay soils which are subject

a

b

5.7 (a) *Viburnum opulus*, a common shrub of lowland woods on rich soils. (b) *Allium ursinum*, the pungent smell of which is unmistakable in woodlands in spring.

to waterlogging; it does not coppice readily but is widely managed as a timber tree.

The tree layer consists almost entirely of *Fagus sylvatica* with or without *Quercus robur*. *Acer campestre* (field maple) and *Fraxinus excelsior* are frequent associates, whilst *Sorbus aria* (whitebeam) and *Prunus avium* (wild cherry) are occasional. A subsidiary tree layer of *Taxus baccata* (yew) or *Ilex aquifolium* forms a continuous stratum underneath the beech in some woods. Where a shrub layer is present the commonest species are *Crataegus monogyna* and *Sambucus nigra* with, occasionally, *Euonymus europaeus* (spindle) and *Corylus avellana*. The climbers *Clematis vitalba* (traveller's joy) and *Hedera helix* are frequent.

The most abundant species of the field layer are *Mercurialis perennis* and *Rubus fruticosus*, in association with *Galium odoratum* (sweet woodruff), *Sanicula europaea* (sanicle) and *Viola riviniana*. Other common species are *Fragaria vesca* (wild strawberry), *Hyacinthoides non-scripta*, *Lamiastrum galeobdolon*, *Anemone nemorosa*, *Arum maculatum* (cuckoo-pint), *Circaea lutetiana* (enchanter's nightshade) and *Carex sylvatica*. Some less common plants characteristic of these beechwoods are *Helleborus viridis* (green

hellebore), *Daphne laureola* (spurge laurel) and the orchid *Cephalanthera damasonium* (white helleborine).

The bryophyte ground layer of beechwoods is discontinuous and most species are restricted to tree bases, the exposed roots of the trees, and the ground in the vicinity of these. The commonest mosses are *Hypnum cupressiforme* on the bases and lower tree trunks and *Ctenidium molluscum* on soil.

5.2.3.3 *Ashwoods on limestone (Fraxino-Brachypodion)*

In the north and west of Britain the extensive limestone of the Lower Carboniferous period forms low hills, escarpments and plateaux, especially in upland areas, and is responsible for the conspicuous landscapes of the Mendip Hills, the southern Pennines of Derbyshire, the northern Pennines of west Yorkshire and north Lancashire, and the Durness limestone of north-west Scotland. Limestone of the same geological period occurs in the Irish Plain. Remnants of natural ashwood and associated calcareous scrub occur in almost all of these areas and also on the smaller limestone outcrops of other geological formations. In common with beech, ash woodland may have expanded as a result of anthropogenic forest clearance. Most woodlands were formerly managed as coppice or coppice-with-standards.

The dominant tree in the woodlands of this alliance is *Fraxinus excelsior*. However, in some areas ash has been felled and replaced with *Acer pseudoplatanus* (sycamore) which, although not a native British tree, grows very well on limestone soils. *Ulmus glabra* (wych elm) may be co-dominant with the ash, but in most woodlands *Fraxinus* is more abundant; *Acer campestre* is a frequent associate, whilst *Quercus robur, Tilia cordata* (small-leaved lime), *Prunus avium* (wild cherry), *P. padus* and *Taxus baccata* are more occasional.

In contrast to oak and beech, Ash does not form a closed canopy and the increased light reaching the ground allows considerable species diversity in the shrub and field layers. This diversity also reflects the favourable nature of limestone soils for the growth of a large number of plants. The commonest shrubs are *Crataegus monogyna* and *Corylus avellana*. *Cornus sanguinea* (dogwood), *Euonymus europaeus, Rhamnus catharticus* (buckthorn), *Ligustrum vulgare* (privet), *Sambucus nigra, Viburnum opulus* and *Rosa* spp. (roses) are also of fairly widespread occurrence. *Viburnum lantana* (wayfaring tree) and the climber *Clematis vitalba* are more or less restricted to southern England.

The seasonal aspect of ashwoods is very pronounced in the field layer with the prevernal plants *Anemone nemorosa, Allium ursinum, Primula vulgaris, Ranunculus ficaria* (lesser celandine), *Mercurialis perennis, Adoxa moschatelina* (moschatel) and *Arum maculatum* flowering before the tree and shrub canopies have fully developed. This period is followed by a sequence of vernal plants: for example, *Silene dioica* (red campion), *Hyacinthoides non-scripta, Viola riviniana* and *Fragaria vesca*. These are succeeded, in turn, by the shade-tolerant aestival plants *Geum urbanum* (wood avens), *Sanicula europaea* and the grasses *Brachypodium sylvaticum* (false brome) and *Melica uniflora* (wood melick).

On sloping ground, subject to periodic water movement, the field layer is dominated by *Deschampsia*

cespitosa, which grows with *Filipendula ulmaria, Geum rivale* (water avens), and *Valeriana officinalis* (valerian), whilst in shady, humid valleys can be found several species of ferns including *Asplenium scolopendrium* (hart's tongue), *A. trichomanes* (maidenhair spleenwort), *A. ruta-muraria* (wall rue), *Polystichum aculeatum* (hard shield fern), *P. setiferum* (soft shield fern), *Dryopteris filix-mas* (male fern) and *Athyrium filix-femina* (lady fern).

The steep sides of many limestone valleys consist of unstable scree and, although the ash trees retain a foothold in the deeper layers of rock and soil, the surface movement of stones prevents the accumulation of surface humus. Very few other plants can establish themselves in this situation. Amongst those that do grow are *Convallaria majalis* (lily-of-the-valley), *Hypericum hirsutum* (hairy St John's wort), *Teucrium scorodonia* (wood sage), *Rubus saxatilis* (stone bramble) and *Geranium lucidum* (shining cranesbill).

The bryophyte layer of limestone ashwoods is very diverse and luxuriant, and includes a number of rare or uncommon species. Although mosses and liverworts colonise the ground, fallen trees, tree stumps and small rocks, large boulders and walls, tree trunks and bases, the rainfall is not normally high enough to support a well-developed epiphytic flora on the upper tree trunks and branches. The commonest species are: on the ground – *Hylocomium splendens, Thuidium tamariscinum, Atrichum undulatum, Plagiomnium undulatum* and *Pseudoscleropodium purum*; on rocks and walls – *Ctenidium molluscum, Homalothecium sericeum, Hypnum cupressiforme, Tortula muralis, Tortella tortuosa* and *Neckera complanata*; tree bases have abundant cover of *Mnium hornum* and *Plagiothecium denticulatum*; and damp valley bottoms and sheltered cliff overhangs favour the growth of the mosses *Neckera crispa, Fissidens* spp. and *Rhizomnium punctatum* and the thallose liverworts *Conocephalum conicum* and *Preissia quadrata*.

5.2.3.4 *Deciduous mixed woodlands (Carpinion betuli)*
In central and southern England a number of forest trees form mixed woodlands on fertile brown earth soils. These mixed woods resemble the oak-hornbeam forests of central and eastern Europe but have been extensively managed for timber and coppice.

The dominant trees are *Quercus robur, Fraxinus excelsior, Acer campestre, Betula pendula, B. pubescens* and, less frequently, *Tilia cordata* (small-leaved lime)

5.8 Inside a beechwood on the chalk of southern England (Fagion sylvaticae). Note the poorly developed ground layer owing to exclusion of light in summer by the dense tree canopy and the copious litter produced each autumn at leaf fall.

5.9 Fraxino-Brachypodion: part of Rassall Wood, the most northerly ashwood in the British Isles on the Durness limestone of Wester Ross.

(especially in central England), *Carpinus betulus* (horn-beam) (particularly in south-east England and East Anglia, often on poorly drained soils) and *Populus tremula* (aspen). The subsidiary tree layer usually contains *Malus sylvestris* (crab apple).

Coppicing has encouraged the growth of *Corylus avellana*, which, as a result, assumes dominance in the shrub layer. However, many other species may also occur, including *Crataegus monogyna, C. oxycanthoides* (midland howthorn), *S. caprea* (goat willow), *Cornus sanguinea*, *Euonymus europaeus*, *Prunus spinosa* (black-thorn), *Rosa* spp., *Sambucus nigra* and *Viburnum opulus*. *Hedera helix* and *Lonicera periclymenum* are woody climbers in the understorey.

The composition of the field layer is influenced by woodland management (if any) and the degree of soil wetness. In the early years of the coppicing cycle, much light reaches the ground and the field layer is very well developed: *Anemone nemorosa*, *Hyacinthoides non-scripta*, *Mercurialis perennis* and, in East Anglian woods, *Primula elatior* (oxlip) are characteristic early-flowering woodland plants. Additional species include *Rubus fruticosus* and the grasses *Deschampsia cespitosa*, *Brachypodium sylvaticum* and *Holcus mollis*. On wet soils communities of *Angelica sylvestris* (wild angelica), *Filipendula ulmaria*, *Urtica dioica* (common nettle), *Ranunculus repens*, *Cirsium palustre* (marsh thistle), *Geum rivale* and *Poa trivialis* (rough meadow-grass) can be found.

The bryophyte cover on the ground varies depending on the humidity and on the nature of the tree canopy: in very shady woods with a deep accumulation of tree litter the bryophyte layer is poorly represented. The commonest bryophyte is the moss *Atrichum undulatum*, which grows on soil. Also occurring, where conditions permit, are *Pseudoscleropodium purum*, *Brachythecium rutabulum*, *Eurhynchium praelongum*, *Fissidens taxifolius*, *Hypnum cupressiforme*, *Mnium hornum* and the liverwort *Lophocolea bidentata*.

5.10 Mixed deciduous woods (*Carpinion betuli*): the main ride at Kirton Wood, Nottinghamshire, an oak/ash/elm wood on Keuper marl (most of the elm has now died owing to Dutch elm disease).

5.3 Grasslands

Most British grasslands have been derived from woodland and are maintained by anthropogenic activities. Those semi-natural grasslands that are maintained solely by grazing and/or mowing but which are neither ploughed nor fertilised are considered in this section. Grassland which is grazed more or less continuously throughout the year is termed 'pasture'. 'Meadow' land is cut for hay during early summer followed by 'aftermath' grazing by stock until the autumn. Pasture includes the 'rough' or 'hill grazings' on the uplands of western and northern Britain, as well as the more floristically diverse chalk and limestone grasslands of central and southern England. Hay meadows, many of which are flood meadows on damp, fertile soils beside watercourses, are much less widespread than they used to be owing to drainage and pasture improvement.

Grazing animals can influence the relative abundance of plant species in a habitat (Harper, 1977). In particular, palatable grasses are eaten at the expense of less-favoured coarse species; in time the latter come to dominate the sward. Trampling and deposition of dung and urine are also important. Hill grazings on acid soils are often dominated by *Nardus stricta* (mat-grass), a tough wiry grass, which has expanded as a result of intense sheep grazing. On chalk grasslands *Brachypodium pinnatum* (tor-grass) and *Cirsium acaulon* (dwarf thistle) are invasive, unpalatable species, which increase under heavy grazing pressure. Prior to the outbreak of the myxomatosis virus in the late 1950s, the feral rabbit population also exerted a strong biotic influence on grassland sward composition. The floristic diversity of chalk grassland is in part due to grazing by rabbits, which selectively defoliate potentially dominant grasses. When the rabbit population dropped so dramatically in the early 1960s the reduced grazing pressure permitted succession to less diverse, tall grassland and scrub.

5.3.1 *Acid grasslands (Nardo-Callunetea)*

These grasslands are in the order Nardetalia of the class Nardo-Callunetea, which occurs throughout Europe eastwards to Siberia. It is best represented in Atlantic, sub-Atlantic and sub-continental regions, but is absent from the more continental areas of central Europe. Associations of lowland heaths are also included within this class, in the order Calluno-Ulicetalia. These are discussed in section 5.4.

The commonest acid grassland associations in Britain are in the alliance Nardo-Galion saxatilis. All are similar in species composition except that *Nardus stricta* (mat-grass) predominates in some whilst *Agrostis canina* and *A. capillaris* (bent-grasses), *Festuca ovina* (sheep's fescue) and *Anthoxanthum odoratum* (sweet vernal-grass) are abundant in others.

Grassland dominated by *Nardus stricta* is more widely distributed throughout upland Britain than bent-fescue grassland and is particularly common in the central Highlands and southern uplands of Scotland, the English Lake District, the Pennines and Wales. This is the predominant grassland type of heavily grazed upland pastures on shallow, podzolised soils overlying hard, siliceous rocks. The characteristic species, apart from *Nardus stricta*, are very similar to those of the bent-fescue grassland, although *Deschampsia flexuosa* (wavy hair-grass) often occurs in abundance. *Juncus squarrosus* (heath rush) is commonly associated with *Nardus stricta*, especially in damp places, and frequently replaces the latter owing to its invasive, spreading growth form.

Regional variants of the *Nardus* grasslands can be distinguished which reflect either excessive grazing or nutrient enrichment of the soil in limestone or flushed areas. The heavily grazed grasslands of the southern Pennines are extremely species-poor, in some cases containing as few as five plant species. Where periodic flushing by water enriches the soil a more diverse grassland results, with the addition of *Carex echinata* (star sedge), *C. panicea* (carnation sedge), *C. pulicaris* (flea sedge) and *Ranunculus acris* (meadow buttercup). There is also a larger and more varied bryophyte component. At higher altitudes these acid grasslands are replaced by sub-alpine heaths (section 9.4.2).

The bent-fescue grasslands are of widespread occurrence in upland Britain on well-drained, alluvial soils beside streams and on hill slopes. These soils are usually well-drained, nutrient-poor podzols with a surface layer of moder humus. In addition to the dominant grasses the most frequent plants are *Galium saxatile* (heath bedstraw), *Potentilla erecta* (tormentil), *Viola riviniana* (common dog violet) and the mosses *Rhytidiadelphus squarrosus* and *Hylocomnium splendens*, with the

5.11 View of *Nardus stricta* grassland (Nardo-Callunetea) in mid-Wales.

occasional occurrence of *Carex pilulifera* (pill sedge), *Luzula campestris* (field woodrush), *L. multiflora* (heath woodrush) and the mosses *Pleurozium schreberi* and *Thuidium tamariscinum*. Some of the mosses indicate a previous forest cover.

5.3.2   *Neutral grasslands (Molinio-Arrhenatheretea)*

This class contains the associations of hay meadows and lowland permanent pastures, and their adjacent wayside and field margins, throughout the British Isles. In Europe they extend from the Eurosiberian region to the Mediterranean wherever the ground water table is fairly high. These grasslands are anthropogenic with a preponderance of caespitose (tufted) grasses and are important economically in providing fodder for livestock, either directly through grazing or indirectly as hay or silage. They are a secondary development from land that was formerly wooded and are maintained by man's agricultural practices. The floristic composition is greatly influenced by the frequency and intensity of these operations. Subsidiary, but equally important factors which produce differences in species composition are moisture, and nutrient and organic contents of the soil.

On the basis of the soil moisture levels the class is divided into two orders: Molinietalia – wet meadow associations, and Arrhenatheretalia – damp to dry meadow associations. In both cases the vegetation develops over a range of soil types, although not on extremes of acidity or alkalinity.

Characteristic species of both orders comprise a large number of grasses including *Alopecurus pratensis* (meadow foxtail), *Festuca pratensis* (meadow fescue), *F. rubra* (red fescue), *Holcus lanatus* (yorkshire fog), *Phleum pratense* (timothy), *Poa pratensis* (smooth meadow-grass) and *P. trivialis* (rough meadow-grass). The herbaceous component contains many colourful plants, amongst which are *Cardamine pratensis* (cuckoo flower), *Centaurea nigra* (common knapweed), *Cerastium holosteoides* (common mouse-ear), *Lathyrus pratensis* (meadow vetchling), *Polygonum bistorta* (common bistort), *Ranunculus acris* (meadow buttercup), *Rumex acetosa* (common sorrel), *Trifolium dubium* (lesser trefoil), *T. pratense* (red clover), *Plantago lanceolata* (ribwort plantain), *Prunella vulgaris* (self-heal), *Vicia cracca* (tufted vetch) and *Lychnis flos-cuculi* (ragged robin). Species less common in the British meadow flora than in Europe are *Colchicum autumnale* (meadow saffron), which occurs in southern England, and *Fritillaria meleagris* (fritillary), a very local plant of

southern and central England. Bryophytes are occasionally present, of which the commonest are *Climacium dendroides* and *Rhytidiadelphus squarrosus*.

5.3.2.1   *Periodically wet meadows (Molinietalia)*

Associations of this order develop where the ground water level fluctuates, but is maintained at a high level throughout the year. The resultant soils are gleyed, water permeability is poor and winter flooding (between late autumn and early spring) is a regular feature. Although some of the associations of this order have probably existed for a very long time restricted to the edges of damp woodlands, they have extended their distribution greatly as a result of the drainage of wetlands and deforestation on damp soils. In both cases recolonisation by scrub and woodland species is prevented by mowing or grazing. Occasionally Molinietalia associations take over damp, previously cultivated fields which have been allowed to remain fallow.

The following species are characteristic of the Molinietalia and its three component alliances, Calthion palustris, Filipendulion and Junco-Molinion: *Equisetum palustre* (marsh horsetail), *Deschampsia cespitosa* (tufted hair-grass), *Dactylorhiza majalis* subsp. *praetermissa* (southern marsh orchid), *Sanguisorba officinalis* (great burnet), *Lathyrus pratensis* (meadow vetchling), *Angelica sylvestris* (wild angelica), *Lysimachia vulgaris* (yellow loosestrife), *Achillea ptarmica* (sneezewort), *Cirsium palustre* (marsh thistle), *Thalictrum flavum* (common meadow rue), *Ophioglossum vulgatum* (adder's tongue fern), *Lychnis flos-cuculi* (ragged robin) and *Valeriana dioica* (marsh valerian).

The alliance Calthion palustris contains associations of soils with a high organic matter content and with a water table which is at or just below the surface for most of the year. Apart from *Caltha palustris* (marsh marigold) the characteristic species include other wetland plants: *Lotus uliginosus* (greater bird's foot trefoil), *Carex disticha* (brown sedge), *Scirpus sylvaticus* (wood club-rush), *Lychnis flos-cuculi* and, particularly in northern Britain, *Crepis paludosa* (marsh hawk's beard).

Filipendulion associations grow on humus-rich, damp soils high in nitrogen which are subject to periodic flooding, especially along streams, rivers and drainage channels. Characteristic species include *Hypericum tetrapterum* (square-stalked St John's wort), *Lythrum salicaria* (purple loosestrife), *Stachys palustris* (marsh woundwort), *Eupatorium cannabinum* (hemp agrimony), *Epilobium hirsutum* (great willowherb) and the grasses *Calamagrostis canescens* (purple small-reed) and *Phalaris arundinacea* (reed canary-grass).

Associations of the most acid substrates in the Molinio-Arrhenatheretea are in the alliance Junco-Molinion which often form the boundary between wet, peat-forming associations of the classes Oxycocco-Sphagnetea (section 7.3.2) or Parvocaricetea (section 6.5.6) and terrestrial associations. This alliance is characterised by the presence of *Succisa pratensis* (devil's bit scabious), *Parnassia palustris* (grass of Parnassus), *Danthonia decumbens* (heath-grass) and *Molinia caerulea* (purple moor-grass).

5.3.2.2 *Damp and dry meadows (Arrhenatheretalia)*
These meadows develop throughout Europe from low-land to alpine regions on soils with a moderate moisture content, although the ground water may fluctuate within wide limits at different times of the year. Unlike the associations of the Molinietalia, however, the water does not usually reach the surface and flooding, if it occurs at all, is a very occasional and short-lived phenomenon. The grasslands of this order are secondary and maintained by man, in particular by grazing and mowing. Many have been converted to arable farmland in recent years. They are very important economically for their luxuriant cover of fodder grasses and herbs, especially leguminous plants. Depending on their land use the Arrhenatheretalia can be divided into two alliances: Arrhenatherion elatioris (hay meadows) and Cynosurion cristati (grazed meadows).

Apart from hay meadows, associations of the alliance

a   b

5.12 Plants of damp meadows (Molinietalia):
(a) *Ophioglossum vulgatum*; (b) *Fritillaria meleagris*.

Arrhenatherion elatioris also occur in roadside verges and hedgerow margins. In all cases the species composition is maintained by regular cutting, often several times a year. Characteristic species include the grasses *Dactylis glomerata* (cock's foot), *Festuca rubra* (red fescue) and *Alopecurus pratensis* (meadow foxtail) and the herbaceous plants *Ranunculus acris* (meadow buttercup), *Lathyrus pratensis, Taraxacum officinale* (dandelion), *Bellis perennis* (daisy), *Leucanthemum vulgare* (ox-eye daisy), *Heracleum sphondylium* (hogweed) and *Trifolium* spp. (clovers). In eastern Britain *Saxifraga granulata* (meadow saxifrage) is a local plant of these grasslands.

Associations of the alliance Cynosurion cristati develop on ground that is grazed and not mown, on soils with nutrient and moisture contents similar to the Arrhenatherion associations. The species diversity is low as a result of grazing and trampling and certain plants, for example *Lolium perenne* (perennial rye-grass), *Poa annua* (annual meadow-grass), *Trifolium repens* (white clover) and *Bellis perennis* assume dominance. The grass *Cynosurus cristatus* (crested dog's tail) and *Ranunculus bulbosus* (bulbous buttercup) often occur as characteristic species of the alliance.

5.3.3 Chalk and limestone grasslands (Festuco-Brometea)

The grassland vegetation of chalk and limestone throughout most of central and western Europe belongs to the class Festuco-Brometea. These grasslands occur on shallow or rocky rendzinas which are well drained

a   b

5.13 Additional plants of hay meadows (Arrhenatheretalia) include: (a) *Orchis morio* which is locally abundant in lowland meadows and pastures in England and Wales and (b) *Primula veris* which is usually associated with calcareous soils.

a

b

5.14 Limestone grassland views (Festuco-Brometea): (a) Ravensdale in the Derbyshire 'White' Peak District; (b) limestone pavement in Yorkshire near Malham.

and xerothermic ('warm'). Soil water deficits at certain times of the year encourage the establishment of xeromorphic grasses with narrow, setaceous (bristle-like) leaves, for example *Festuca ovina* and *Koeleria macrantha* (crested hair-grass). Limestone grasslands are one of the most species-rich of all the vegetation types of the British Isles and contain a number of brightly coloured dicotyledonous herbs.

In Britain this vegetation is widely distributed on the chalk of south and south-east England from Kent and the Salisbury Plain to the Yorkshire Wolds and along the Jurassic Limestone escarpment which runs through the Cotswolds to the Lincolnshire Edge. The dry grasslands of the older Carboniferous Mountain Limestone areas are less extensive than those of the chalk but examples can be found in Derbyshire and along the south coast of Wales, in the northern Pennines, the Lake District and on the Magnesian Limestone escarpment of Durham. However, in northern England there is an intermingling in the vegetation of some species more characteristic of the cold-temperate and alpine limestone grassland associations of the class Elyno-Seslerietea (section 9.4.5). Further north still the Festuco-Brometea vegetation declines, presumably because of the absence of suitable calcium-rich soils under the cold and wet climate of northern Britain, which favours leaching and podzolisation.

Brometalia erecti, the order of the Festuco-Brometea which occurs in Britain, contains two alliances – Xerobromion and Mesobromion. The former consists of primary vegetation which is thought to have survived the post-glacial forest maxima of the Atlantic period, whilst the latter comprises secondary associations which established following removal of the primeval forest.

5.3.3.1 *Dry pioneer grasslands of steep rocky slopes (Xerobromion)*

These grasslands are associated with chalk and limestone cliffs and develop on thin rendzina soils. This alliance is widely distributed in central and southern Europe but in Britain it is restricted to small areas of Somerset and Devon. Characteristic species are *Helianthemum apenninum* (white rockrose), the umbellifer *Trinia glauca* (honewort), *Scilla autumnalis* (autumn squill), *Koeleria vallesiana* (crested hair-grass) and *Carex humilis* (dwarf sedge).

5.3.3.2 *Chalk and limestone grasslands (Mesobromion)*

This alliance contains most of the chalk and limestone

grasslands of the British Isles together with some communities of stabilised, calcareous sand dunes. These are widely distributed throughout the chalklands of southern England, especially on the North and South Downs, Salisbury Plain and the Chiltern Hills. However, they are of limited occurrence in the upland limestone areas of Britain, where they are only encountered occasionally on scarp slopes and in a variety of man-made habitats such as golf courses, old quarry workings and roadside verges.

Characteristic and widespread species of this alliance include the grasses *Zerna erecta* (upright brome), *Brachypodium pinnatum* (tor-grass) and *Avenula pratensis* (meadow oat), plus *Helianthemum nummularium* (common rockrose), *Asperula cynanchica* (squinancy wort) and *Cirsium acaule* (dwarf thistle). But also occurring are less common species such as *Anacamptis pyramidalis* (pyramid orchid), *Campanula glomerata* (clustered bellflower), *Picris hieracioides* (hawkweed ox-tongue), *Polygala calcarea* (limestone milkwort), *Pulsatilla vulgaris* (pasque flower), *Thymus pulegioides* (larger wild thyme) and *Thalictrum minus* (lesser meadow rue).

The plant associations of the Mesobromion are subject to varying intensities of grazing pressure and many of the species, for example *Briza media* (quaking grass), *Carex flacca* (glaucous sedge), *Centaurea nigra* (lesser knapweed), *Pimpinella saxifraga* (burnet saxifrage) and *Plantago lanceolata* (ribwort plantain), indicate this aspect.

The Mesobromion can be divided into two sub-alliances: Eu-Mesobromion and Seslerio-Mesobromion.

Southern limestone grasslands (Eu-Mesobromion)

This sub-alliance contains almost all of the lowland associations on chalk and limestone as far north as Derbyshire. Different associations reflect regional variations in soil and climate and sub-alpine species, for example, *Sesleria albicans* (blue sesleria) are absent.

Bryophytes are common, especially *Pseudoscleropodium purum* and *Homalothecium sericeum*. The moss *Neckera crispa*, and the liverworts *Frullania tamarisci* and *Scapania aspera* tolerate wet conditions on north and west-facing slopes at higher altitudes, whilst *Rhytidiadelphus triquetrus* and *Hylocomium splendens* are good indicators of the former climax woodland.

Northern limestone grasslands (Seslerio-Mesobromion)

The associations of this sub-alliance occur in a well-defined zone across northern England and in western

5.15 Plants of limestone grassland (Festuco-Brometea): (a) *Pulsatilla vulgaris*, a very local plant in southern Britain; (b) *Carex ornithopoda* which is of restricted distribution in northern Britain.

5.16   Heather moor burning at Wanlockhead, southern Scotland. Note the patchwork of
burning which has been carried out over several years.

Ireland, forming a phytogeographical link between the
classes Festuco-Brometea of central Europe and most
of England, Wales and Ireland and Elyno-Seslerietea
(section 9.4.5) of northern Europe and some mountainous regions of northern Scotland. These grasslands
are characterised by the predominance of *Sesleria
albicans*. Additional species include *Poterium
sanguisorba* (salad burnet), *Scabiosa columbaria* (small
scabious), *Helianthemum nummularium, Helictotrichon
pratense, Koeleria macrantha* and *Carlina vulgaris*
(carline thistle). The damper nature of this grassland is
indicated by the presence of *Primula farinosa* (bird's-eye
primrose).

5.4   Dry heaths (Nardo-Callunetea)

Dry heath, in common with most grasslands, is a product
of diverted succession. In the absence of grazing and
burning, coniferous or deciduous forest would be the
climatic climax vegetation of most present-day dry
heathlands. An exception, however, is in coastal regions,
where persistent strong winds and salt spray prevent
tree growth.

   In Europe, dry heaths have an oceanic distribution
and occur in appropriate locations in northern France,
Belgium, Holland, northern Germany, south-west

Sweden and Denmark. In Britain they are found in East
Anglia, south and west England, some parts of the
Pennines, and eastern and northern Scotland. In wetter,
western districts, dry heath is replaced by the peat-
forming wet heaths or blanket mires (section 7.3.2).

   Heathland development and maintenance is associated with various management practices, notably sheep
grazing and heather burning. The dominant heathland
plant, *Calluna vulgaris*, is most productive at a juvenile
stage in its development, when abundant young, green
shoots are being produced; older, mature bushes are
more woody, less productive and of lower nutritional
value to grazing animals. The most widely practised
management for maintaining *Calluna* at the most productive stage of its life-cycle is burning on a 10–12 year
rotation. If *Calluna* is burnt at this stage in its growth
it will regenerate rapidly from axillary buds and quickly
re-establish dominance over other plant species to form
almost pure stands. If old, degenerate heather is burnt,
however, re-establishment can only take place from
seeds buried in the surface soil layer – a slow process
which is further retarded by the presence of grazing
animals – and a more varied heath community of other
Ericaceous shrubs, mosses and lichens may precede the
eventual return of species-poor *Calluna* heath.
(Gimingham, 1972).

Many heaths are also managed as grouse moor. The preferred habitat of the red grouse (*Lagopus lagopus scoticus*) is heather moorland and the young shoots of *Calluna vulgaris* make up a substantial proportion of the bird's diet. Grouse moors are also managed by rotational burning on a cycle similar to that for sheep. However, in order to create optimal habitat conditions for the grouse (young shoots for food and tall, mature bushes for cover) the heather is burnt in small patches to ensure that each grouse territory contains stands of various ages.

Dry heath communities are in the alliance Calluno-Genistion, order Calluno-Ulicetalia of the class Nardo-Callunetea.

5.4.1 *Heather-petty whin heaths (Calluno-Genistion)*

Throughout Britain these heaths are typical of acid, podzolised soils (pH 3.5–6.5) formed over fast-draining, nutrient-poor sandstone, sand or gravel. They are occasionally found over strongly leached, calcareous substrates, but once established the acid leaf litter of the ericaceous plants depresses the pH of the soil humus and podzolisation is encouraged.

Characteristic plants of the alliance include *Calluna vulgaris* and *Genista anglica* (petty whin). Two geographical variants of this alliance can be distinguished in Britain: oceanic heaths of western seaboards and northern heaths. The former include all of the lowland heaths of England, Wales and Ireland. The dominant species is *Calluna vulgaris*, which usually forms a dense canopy; *Erica cinerea* is a constant associate. Species growing amongst the heather include: *Ulex gallii* (dwarf furze), especially in regions with a strong maritime influence, e.g. Cornwall, Devon, west Wales and western Ireland; *Ulex minor* (small furze) in south-east England where the oceanic influence is less marked; *Genista anglica* (petty whin) on heaths from southern England to north-eastern Scotland; *Erica vagans* (Cornish heath), which is restricted to the Lizard Peninsula; and *Dabeocia cantabrica* (St Daboec's heath), which only occurs in western Ireland. Additional species include *Ulex europaeus* (gorse), *Cytisus scoparius* (broom), *Vaccinium myrtillus*, *Galium saxatile*, *Polygala vulgaris* (common milkwort), *Blechnum spicant* and *Cuscuta epithymum* (common dodder), the last of which is parasitic on *Calluna vulgaris* and *Ulex* spp., plus the grasses *Danthonia decumbens* (heath-grass), *Deschampsia flexuosa*, *Festuca ovina* and, particularly on western heaths, *Agrostis curtisii* (bristle bent). Mosses include *Dicranum scoparium*, *Hypnum cupressiforme*, *Pleurozium schreberi*, *Polytrichum juniperinum* and *Leucobryum glaucum*. Lichens of the genus *Cladonia* also frequently occur.

a   b

5.17 Dry heath vegetation (Nardo-Callunetea): (a) *Calluna vulgaris* and *Erica cinerea*; (b) *Dabeocia cantabrica*.

The northern heaths (often assigned to a separate alliance, the Empetrion nigri) are found on peaty podzols in northern Britain where *Calluna vulgaris* grows with *Vaccinium myrtillus, V. vitis-idaea* and *Empetrum nigrum*. In the most northerly heaths *Arctostaphylos uva-ursi* (bearberry) assumes local dominance. Additional species include *Ulex europaeus, Erica cinerea,* *Pyrola* spp. (wintergreens) and the grasses *Deschampsia flexuosa* and *Nardus stricta*. The species of the moss layer are similar to those of the oceanic heaths with the addition of *Hylocomium splendens* and *Racomitrium lanuginosum*. Northern heaths grade into montane heaths (section 9.4.2.2) in upland regions.

Chapter 6

Freshwater aquatic and swamp vegetation

Aquatic and swamp plant associations are amongst the most natural vegetation to be found anywhere in the world. They have developed independently of man wherever the water table is above, at, or near the surface. They include running waters, standing waters, fens, swamps and marshes. The range of ecological variation within these is considerable and largely results from differences in water chemistry, depth and flow rate. However, not all habitat conditions produced in this way favour plant growth. For example, aquatic vegetation is best developed in small ponds and lakes, slow-flowing rivers and the sheltered bays of larger, deeper lakes. Wave action in lakes and water currents in rivers limit the establishment of bottom rooted aquatic plants and carry away free-floating plants. In aquatic associations the plants live entirely within or on the surface of the water body; flowers are usually emergent. In swamp associations only a small part of the plant is normally under water (the roots and the lower parts of leaves and stem), whilst most of the photosynthetic biomass extends above the water surface. In addition, some communities are amphibious and can tolerate periodic flooding.

Wetlands are dynamic systems which form part of the hydroseral succession from open water to dry land (section 6.4). As a result of agricultural practices (grazing, mowing and removal of plants for bedding or thatching) over many centuries, this succession has been arrested at different stages, e.g. reed swamp or fen meadow. However, man has also reduced the area of aquatic and wetland habitat through drainage of fens and swamps and canalisation of rivers to prevent flooding. This has led to habitat loss and a reduction in species diversity. In more recent years many lowland rivers have been polluted by agricultural, domestic and industrial wastes which have resulted in large increases in the concentrations of certain chemical elements, in particular nitrates, phosphates and toxic compounds. These have further impoverished the plant and animal communities. On the other hand, new wetland habitats have been created in the form of gravel and sand pits, reservoirs, canals and drainage ditches. Many of these support aquatic and swamp communities which, although less diverse than their natural counterparts, are floristically similar.

6.1   Running waters

These are rivers, streams, drainage channels and springs, the main feature of which is a uni-directional water movement. Rivers may be described as passing through a series of developmental stages, from youth through maturity to old age (Fig. 6.1). Youthful streams and rivers have a steep gradient and are fast-flowing. They occupy steep-sided, V-shaped valleys which are under-going rapid erosion and corrasion. Mature, middle-course rivers flow over a shallower gradient in valleys broadened by erosion and in which the volume of water is greater than at the youthful stage, having been supplemented by run-off from the land and tributaries. However, the flow rate is less and they have a tendency to meander across their valleys. In their lower courses rivers flow sluggishly down very slight gradients meandering through broad, flat valleys in which they occasionally form ox-bow lakes. During periods of high water, rivers may overflow their banks and flood surrounding land (the flood plain).

The main effect of water movement is on the size of the mineral particles carried along in suspension or deposited on the river or stream bed. In fast-flowing, upland watercourses small particles of sand, silt and peat are carried away in suspension, whilst heavier gravel and small stones may be rolled along causing corrasion of the bed and deepening of the channel. Transport of

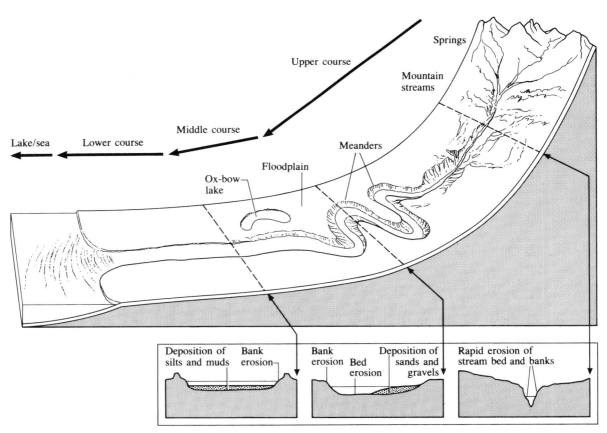

6.1 Stylised development of a river from source to sea (redrawn from Bar, 1976).

large stones requires a strong current, so that as the river gradient decreases these are redeposited. However, the smaller, suspended particles are carried greater distances and redeposited in lowland rivers where the flow rate is slower. Consequently, youthful rivers and streams have rocky beds, with only small accumulations of sand and gravel in sheltered stretches. This is in contrast to the sandy and gravelly beds of mature rivers, and the silts and muds of old-age rivers and man-made dykes or ditches. (In the latter, water flow may be brought almost to a standstill in summer as a result of silt accumulation.) In medium- and slow-flowing rivers the water is deeper and the current stronger in midstream, rather than near the banks; however, where the river bends, a faster flow occurs towards the outside of the curve. These differing current conditions produce an uneven sedimentation of particles on the river bed. Suspended solids affect the clarity of running waters; usually the turbidity of rivers in old age is high owing to fine silt and detritus held in suspension.

The nature of river sediments determines which plants can grow on or in them. Large boulders provide a habitat for only a few attached algae, liverworts and mosses, whereas vascular plants are able to root in finer sediments, although these may vary in depth and are normally unconsolidated and unstable. Current velocities also influence the vegetation composition of running waters. Plants have to maintain their position to avoid being swept downstream and also they must withstand the effects of scouring by materials carried in suspension. (Haslam, 1978).

Upland rivers usually have lower concentrations of all dissolved chemical elements (with the exception of oxygen) than those of lowland regions, which receive large inputs of nutrients, and organic material, from their more extensive catchments. Although the concentration of dissolved oxygen is high in fast-flowing waters as a result of turbulence, incorporation of this gas from the air into slow-flowing rivers is less, and a greater proportion is derived from the photosynthetic activities of submerged green plants. In mature rivers dissolved oxygen levels show considerable seasonal and diurnal variation.

6.1.1 *Springs*

Springs are a special category of running waters in which water temperature and chemical composition vary little throughout the year. The size and location of springs is dependent on the local topography and geology: in regions of porous rock (e.g. sandstones and chalks) water percolates downwards through the rock until it reaches an impermeable stratum, along which it flows to the land surface to emerge as a spring, or the water may re-emerge where the land surface falls below the level of the local water table, e.g. in a valley. Alternatively, water can move through rock fissures, to emerge again at the land surface some way below the level of intake. In limestone areas solution of the rock by percolating rainwater produces an underground network of drainage channels through which the water flows (often considerable distances) before emerging at the surface at a lower altitude.

Downslope from a spring-head the waters may converge to form a stream or, alternatively, spread out over the surface as a wet 'flush', which receives constant mineral enrichment. The percentage plant cover and floristic composition of different springs and flushes is affected by the water flow rate and chemical composition.

6.2 Standing waters

Impoundment of water by natural and man-made obstructions creates areas of standing water: ponds, lakes and reservoirs of varying size. In contrast to the turbulence of running waters, which move downhill under gravity, most movements in standing waters are induced by wave action, the magnitude of which varies with surface area and depth of the water body. Lakes and reservoirs are usually less aerated than rivers. At depth, oxygen levels can become depleted, particularly during summer months, when, as a result of temperature stratification between the warmer, upper (epilimnion) and lower, cooler (hypolimnion) waters, mixing of the layers is prevented. With the onset of winter the surface waters cool, the temperature gradient within the lake decreases and mixing of the two layers occurs, with the result that oxygen levels are raised throughout the lake. (Maitland, 1978).

The flora of the littoral zone of lakes (marginal shallows less than 1 m deep) is greatly influenced by wave action. In sheltered bays and margins with small stones, sand and silt on the bottom, submerged rosette-forming vascular species predominate, and tall emergent species are restricted to the shallower edges. In contrast, the large boulders and bare rocks of wind-exposed shores support a meagre vegetation, mainly a film of

6.2 Lakes of different trophic status:
(a) oligotrophic loch, Diabaig, Wester Ross;
(b) upland marl lake, Sunbiggin Tarn, Cumbria;
(c) dystrophic lochans, Dubh Lochs of Munsary, Caithness.

diatoms, desmids and filamentous algae, on which aquatic bryophytes may grow.

6.3   Trophic status of waters

Waters can be classified according to their primary productivity, i.e. trophic status. This is determined by the chemical composition of the water which, in turn, is influenced by: the geology of the region, through the soils and sediments that are derived from the parent rock; the area of the catchment supplying the water; and the surrounding land use, especially the intensity of inorganic fertiliser application. Most waters can be assigned to one of the following categories: (1) oligotrophic, (2) eutrophic, (3) mesotrophic, (4) dystrophic, (5) marl, or (6) saline (including brackish) (Table 6.1).

6.3.1   *Oligotrophic waters*

In Britain, oligotrophic waters are found mainly in upland areas where high rainfall and hard, acidic rocks provide a plentiful supply of nutrient-poor water. These lakes, lochs, tarns, streams and rivers are a characteristic feature of elevated situations in west and north Wales, the English Lake District, the Scottish Highlands and south-west Scotland (Fig. 6.2).

Light can penetrate to considerable depths in oligotrophic waters, enabling plankton and aquatic macrophytes to photosynthesise well below the surface. However, primary productivity is low since oligotrophic waters contain very low concentrations of dissolved elements, especially nitrogen and phosphorus, which are mainly bound up in the small amount of organic sediment present. Low nutrient levels, combined with the high concentrations of dissolved oxygen, ensure almost complete decomposition of the annual organic production.

6.3.2   *Eutrophic waters*

In contrast to oligotrophic waters, the input of nutrients to eutrophic ones is considerable, resulting in a high primary productivity. Nitrogen and phosphorus are present in sufficient concentration to sustain a large biomass of producer and consumer organisms and, as a result, plankton population densities can increase rapidly during spring and early summer and blooms are frequent. However, low concentrations of dissolved oxygen, especially in late summer, can result in reduced organic matter decomposition, which may lead to oxygen starvation and, in extreme circumstances, death of animal populations. The excess annual biomass which cannot be recycled by decomposers accumulates as sediment on the bottom of eutrophic lakes, whereas in eutrophic rivers excess organic production is transported downstream. Although the surface of eutrophic lakes may remain oxygenated through contact with the atmosphere, the lower layers are organic, anaerobic, and contain high levels of toxic hydrogen sulphide.

The input of organic material of small particle size, much of which remains in suspension in the water, and

Table 6.1 Characteristics of waters of different trophic status (from Ratcliffe, 1977).

Nutrient status	Alkalinity (ppm C_aCO_3)	Winter pH	Water colour	Productivity
Dystrophic	0–2	<6.0	brown, peaty	very low, limited by lack of nutrients and light
Oligotrophic	0–10	6–7	clear	low, limited by lack of nutrients
Mesotrophic	10–30	*c.* 7.0	often greenish	moderate to high
Eutrophic	>30	>7.0	usually green	high
Marl	>100	>7.4	very clear	very low
Brackish	(cond >500 μS mainly NaCl)	variable	clear	variable

the high productivity of phytoplankton and algae, pre-vents light from penetrating very far below the surface of eutrophic lakes and rivers. Primary production of free-floating autotrophs is, therefore, confined to a very thin surface layer, and aquatic macrophytes are restricted to shallow water.

In shallow eutrophic lakes, mineral sediments are rapidly colonised by macrophytic vegetation, the growth of which may be luxuriant and assists in hydroseral infilling (section 6.4). Lakes and rivers in lowland areas are more eutrophic than their upland counterparts and although large eutrophic lakes were formerly a feature of much of the English landscape in fenland districts of the south and east of the country, these have been extensively drained. The majority of lowland eutrophic waters are now artificial water bodies such as canals, flooded gravel pits and reservoirs. The only eutrophic lakes which occur in the uplands are in limestone areas where they have been formed in kettle-holes or morraine-blocked valleys (Fig. 6.2).

6.3.3   *Mesotrophic waters*

Oligotrophy and eutrophy represent the opposite ends of the spectrum of primary productivity in water bodies, determined by major differences in water chemistry. However, lakes and rivers may have a trophic status between these two extremes, a range of variation which is encompassed by the term 'mesotrophic'. The differences between mesotrophic waters and either oligotrophic or eutrophic ones are indistinct. In Britain, mesotrophic lakes and rivers are found on the fringes of upland areas where water running off acidic, nutrient-poor rocks is enriched by water from the lower and more extensive catchments of the foothills.

6.3.4   *Dystrophic waters*

These waters have a high content of colloidal organic material (humus) and, although the concentration of hydrogen ions in the water is high, that of all other dissolved elements is very low. The large amount of organic matter in suspension in the water, which imparts to it a dark-brown colouration, reduces the dissolved oxygen concentration to very low levels, except at the surface where wind-induced turbulence causes a small amount of oxygen to dissolve. This promotes some organic decomposition and the release of nitrogen and phosphorus, which may be sufficient to support

growth of algae on the water surface in summer. Algal photosynthesis further increases the dissolved oxygen content of the water, which subsequently sustains an augmented rate of organic matter decomposition, giving rise to a very fine-particled organic sediment called 'Dy' (Swedish), which accumulates on the bottom.

Since dystrophic lakes, pools and streams frequently occur in, or flow from, peat bogs, their distribution in Britain follows that of ombrotrophic mires (section 7.2). Dystrophic standing waters range from intermediate-sized lakes to small bog pools. The former are oligotrophic lakes which receive drainage water high in humic colloids from peat-covered catchments at higher altitudes. The origin of dystrophic bog pools, however, is less clear. In lowland areas they may represent the last remains of former lakes which have infilled during hydroseral successional changes; in upland regions they may have developed as a result of degeneration and erosion of the peat surface of senescing blanket mires (Fig. 6.2; see also section 7.1.6).

6.3.5   *Marl lakes*

Marl is a precipitate of calcium carbonate which is deposited on the bottom of lakes in limestone districts where calcium is the predominant cation in the water. This process is induced by the removal of carbon dioxide from water by photosynthetic organisms. Associated with marl accumulation is the removal of phosphorus from solution as insoluble calcium phosphate, and this lack of phosphorus, coupled with the low levels of dissolved carbon dioxide, reduces the productivity of producer organisms to very low levels. Consequently, marl lakes are both very clear and unproductive, aspects in which they resemble oligotrophic waters. Marl lakes are confined to a few limestone and chalk areas in upland and lowland Britain.

6.3.6   *Saline and brackish waters*

Apart from some artificial lakes in the Midlands of England in the vicinity of salt workings, most of these waters are coastal. They include: brackish waters influenced by sea-spray; tidal channels where fresh and sea water mix; and salt-water, tidal marshes. All saline waters are characteristically high in available elements of marine origin, especially sodium, the concentrations of which depend on the extent of freshwater dilution and degree of daily exposure to tidal movements. In

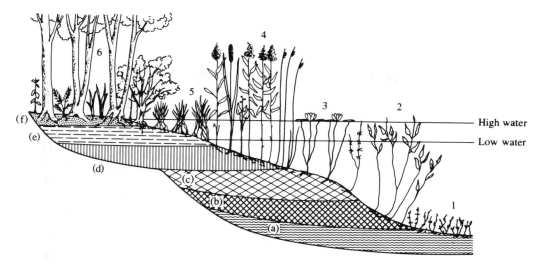

6.3 Terrestrialisation of eutrophic standing water through the succession of plant communities to the climax woodland stage (after Firbas, 1952 in Ellenberg, 1963).
1 *Chara* spp. (submerged algae)
2 Rooted submerged and floating species (e.g. *Potamogeton, Myriophyllum*)
3 Rooted floating-leaved species (e.g. *Nymphaea*)
4 Emergent reeds and grasses (e.g. *Phragmites, Typha, Schoenoplectus*)
5 Tall sedge swamp, emergent (e.g. *Carex elata*)
6 Alder woodland (*Alnus glutinosa*)
The difference between low and high water is less than 1 m. The sediments include:
   (a) Gyttja and lake chalks
(b) (c) Muds with plants remains
   (d) Reed peat
   (e) Sedge peat
   (f) Wood peat

Britain, this is a relatively small group of water bodies, extremely variable in size and nutrient content. As salinity increases with proximity to the sea the vegetation increasingly includes species characteristic of the marine littoral zone (section 8.1).

6.4 The hydrosere

The hydrosere is a continuum of vegetation types which replace each other as succession proceeds from open water to forested mire. These changes may be initiated in waters of any trophic status from oligotrophic to eutrophic. However, the rate of infilling of the water body by accumulating sediments, and the length of time that each vegetation type survives, is closely related to water chemistry and depth. Shallow, eutrophic lakes complete all stages of the hydrosere more quickly than either deep eutrophic lakes or oligotrophic ones, whatever their depth. Deep oligotrophic lakes, because of their

low nutrient content and efficient decomposition cycle, may never infill. (Moore & Bellamy, 1973).

Although stages in the hydrosere can be observed in most fresh waters, it is in enclosed ponds and lakes that the succession of plant communities from open water to dry land is best expressed. Open waters contain free-floating plants – phytoplankton, algae and members of the Lemnaceae (the smallest British flowering plants). In shallow water much of the surface is occupied by large, floating leaves of bottom rooted, aquatic angiosperms, e.g. pondweeds and water lilies. In contrast, plants of the shallower perimeter are emergent (Fig. 6.3).

Water depth decreases over time as a result of infilling by sediments of either external (allochthonous) or internal (autochthonous) origin. The former are mainly mineral inwash materials transported by streams and rivers from the surrounding catchment, whilst the latter comprise organic matter formed *in situ* by plants and animals growing in and around the water body. In eutrophic waters this organic material does not undergo complete decomposition and, with time, a progressive accumulation and compaction of the dead remains takes place. This surplus, together with allochthonous silts, accumulates on the bottom, gradually reducing the water depth. Marginal areas become progressively shallower and the area of open water diminishes. Initially this 'terrestrialisation' is manifested in small areas which are isolated above the water surface during periods of low water level, but eventually, as a result of continued accrual of sediments, large expanses of vegetation are maintained above the ground water influence for longer and longer periods each year.

Throughout these later successional changes plant productivity remains high until the lake is almost completely filled in. The accumulating peat eventually occupies the entire lake basin and further accrual raises the surface and its vegetation above the influence of the mineral ground water. The source of nutrients for plant growth during the terminal stages of the hydrosere is from precipitation only, which is naturally acid and low in dissolved chemical elements. This transition from a ground- to a rainwater supply is accompanied by colonisation of the surface by acidophilous plant communities (Ch. 7), with a consequent decrease in above-ground productivity.

6.5   The vegetation of freshwater lakes, rivers and swamps

Throughout this book the term 'mire' is applied, in a general sense, to those stages of the hydrosere where there is a high water table, suppression of organic matter decomposition and hence the 'potential' for partially decomposed or undecomposed material to accumulate. This process can be maintained under two distinct regimes of water supply: moving, or seeping, 'geogenous' ground water, which is enriched with dissolved nutrients derived from rocks and soils; and 'ombrogenous' water, which comes entirely from aerial precipitation.

The terms 'ombrotrophic' (Greek – nourished from above) and 'rheotrophic' (Greek – nourished by the flow) are used to distinguish between mires formed under the varying influences of ground water and precipitation. Ombrotrophic mires ('bogs') receive ombrogenous water only and, as a consequence, are acid and nutrient-deficient. A range of rheotrophic mires develop in

response to differences in the chemistry, flow rate and depth of geogenous waters. These form a continuum of mire types from calcareous, nutrient-rich habitats (eutrophic) through to acid, nutrient-deficient ones (oligotrophic) in which there is incipient ombrotrophic peat development. The term 'transition mire' is often used to describe the intermediate stages of the hydrosere.

Plant communities of rheotrophic or transition mires dominated by emergent species are frequently referred to as 'fen'. 'Swamp', or 'reed-swamp', are types of fen in which tall, often mono-dominant stands of emergent grasses and sedges colonise organic or mineral substrates around open water. 'Marsh' is the wet mineral-ground margin of freshwater swamp or fen, particularly where waterlogging occurs during the winter months only. 'Carr' and 'fen carr' are the low-growing scrubby woodland which often colonises fen peat as it dries out following natural changes in water levels or artificial drainage. (Gore, 1983).

Freshwater plants differ from those of terrestrial habitats in several important respects. Plants of fast-flowing waters have strong adventitious roots or rhizomes which penetrate between stones and gravel on the river bed, forming a strong anchor. Leaves and stems are flexible and streamlined to reduce resistance to water flow, so that long, narrow stems and petioles and strap-shaped or finely divided leaves are common. Some species exhibit heterophylly, e.g. *Sagittaria sagittifolia* (arrowhead), which produces two, and sometimes three, types of leaf on the same plant – linear submerged leaves, lanceolate to ovate floating leaves and broad arrow-shaped aerial leaves (Haslam, 1978). The thin leaves of submerged plants have neither cuticle nor stomata and they exchange gases directly with the water. Although most aquatic plants can absorb dissolved carbon dioxide, some plants of calcareous waters can also utilise bicarbonate ions as a carbon source for photosynthesis. The absence of bicarbonate ions from oligotrophic waters may be a factor limiting plant growth.

Submerged plants receive less light than emergent species because of: reflection by the water surface (between 5% and 25% of the incoming solar radiation may be lost in this way); attenuation in the water (absorbence by dissolved substances and scatter by particles in suspension, including algae and plankton); and shading by marginal vegetation. There is usually insufficient light for plants to grow at depths greater than 10 m even in clear waters. In turbid waters, however, this depth is greatly reduced and plant growth is restricted to the uppermost layer of only a few centimetres. Some aquatic plants have high light requirements, e.g. *Ranunculus* spp. (water crowfoots), whilst other species have evolved 'shade forms' in which the rate of respiration is depressed. This enables them to survive despite reduced rates of photosynthesis in low light conditions. Some aquatic species are capable of growing in only 1% of the light available at the water surface. (Fitter and Hay, 1981).

Aquatic and wetland macrophytes show a variety of life forms, and the following groups can be distinguished:

1. Submerged plants, which are rooted in sediments or attached to other underwater materials. Their entire photosynthetic biomass (leaves and green stems) is below the surface of the water, but the reproductive organs may be aerial, floating or submerged. Examples include aquatic bryophytes, *Elodea canadensis* (Canadian pondweed) and *Potamogeton* spp. (pondweeds).
2. Floating-leaved plants, which may be rooted or free-floating, and bear some, if not all, of their photosynthetic biomass at the water surface. Reproductive organs are also floating or completely aerial. Examples are *Nuphar* and *Nymphaea* spp. (water lilies); free-floating plants include *Lemna* spp. (duckweeds).
3. Emergent plants which are rooted underwater but produce most of their photosynthetic biomass and reproductive organs above the water surface include many of the species of fen and reed-swamp, e.g. *Phragmites australis* (common reed) and *Carex* spp. (sedges).

The vegetation of rheotrophic and transition mires is considered in this chapter, along with other plant communities that are strongly influenced by moving or seeping ground water. Ombrotrophic mires and their vegetation are described in Chapter 7.

6.5.1 *Free-floating plants (Lemnetea)*

The single order Lemnetalia of this class contains all the surface, free-floating associations of sheltered, eutrophic, mesotrophic and brackish waters throughout the British Isles, excluding the far north of Scotland. The character species of the order is *Lemna polyrhiza* (greater duckweed), which commonly occurs with *L. minor* (common duckweed) and *L. trisulca* (ivy-leaved duckweed). The small, floating thalli of these plants

form bright green carpets on the surface of still or stagnant waters. However, in large lakes with a strong wave action, and in rivers, they are restricted to backwaters and sheltered bays. Species of this class can also exploit temporary habitats, e.g. ephemeral pools and drainage ditches which dry out in summer. *Wolffia arrhiza* (rootless duckweed), the smallest British flowering plant, occurs rarely in association with *Lemna* spp. in still waters in southern England, from Somerset to Kent. In shallow eutrophic waters (up to 60 cm) species of *Lemna* may also be found growing with the small, free-floating, thallose liverworts *Riccia fluitans* and, less commonly, *Ricciocarpus natans*.

6.5.2 *Deep-water communities (Charetea and Potametea)*

Associations of the class Charetea, dominated by large algae of the Charophyta ('stoneworts'), occur in many lakes throughout the British Isles. Depending on the clarity of the water, these can grow to a depth of 5 m and often form dense swards on the bottom of lakes, where they are attached to the substrate by rhizoids. Species of *Chara*, many of which are encrusted with external deposits of calcium carbonate, occur in eutrophic and marl lakes; members of the genus *Nitella*, which are non-encrusted, also occur in waters with a fairly high calcium content.

Often associated with species of the Charetea, in shallower waters (less than 3 m) are communities of bottom rooted angiosperms of the class Potametea, which form floating-leaved or submerged plant associations in mesotrophic, eutrophic and brackish waters. These plants, which are pioneer species of the initial stages of the hydrosere, include *Myriophyllum spicatum* (spiked water milfoil), *M. alterniflorum* (alternate water milfoil), *Elodea canadensis*, *Potamogeton pusillus* (lesser pondweed), *P. alpinus* (red pondweed) and *Sparganium minimum* (least bur-reed).

The Potametea contains three orders:

(i)   Magnopotametalia – associations of tall, submerged pondweeds and surface-floating plants;
(ii) Parvopotametalia – small, mainly submerged aquatic plants; and
(iii) Luronio-Potametalia – which contains the few submerged associations of shallow, acid waters.

6.5.2.1 *Large pondweeds (Magnopotametalia)*
Communities dominated by large pondweeds occur

6.4   Floating-leaved plant communities (Magnopotametalia): *Nuphar lutea*.

throughout the British Isles in eutrophic and meso-trophic, slow-flowing and standing waters usually more than 1 m deep. There are two alliances: Magnopotamion and Nymphaeion. Associations of the former are dominated by submerged pondweeds, for example, *Potamogeton lucens* (shining pondweed), which has a preference for calcareous waters with base-rich, inorganic substrates; and *P. perfoliatus* (perfoliate pond-weed) and *P. pectinatus* (fennel pondweed), which occur in brackish coastal waters and inland, lowland waters where the substrates are only moderately organic. The alliance Nymphaeion contains the associations of large floating-leaved plants of sheltered waters. Characteristic species are *Nymphoides peltata* (fringed water lily), a locally occurring plant of lowland alkaline waters over mineral, often clayey substrates, and the commoner *Nuphar lutea* (yellow water lily) and *Nymphaea alba* (white water lily), which root in organic sediments. In very shallow waters plants of the other two orders within the class may appear as co-dominants. These communities are usually succeeded by those of the Phragmitetea (section 6.5.4).

6.5.2.2 *Small pondweeds (Parvopotametalia)*
Associations of this order occur in shallow meso- to eutrophic or brackish waters less than 1 m deep and are dominated by small aquatic species: the pondweeds

Potamogeton crispus (curled pondweed), *P. pusillus, P. berchtoldii* (small pondweed), *Elodea canadensis* and *E. nutallii* (pondweeds introduced from Canada which have become naturalised in Britain); water milfoils (*Myriophyllum* spp.); and water crowfoots (*Ranunculus* subgenus *Batrachium*). Of the three alliances, Parvopotamion, Hydrocharition and Callitricho-Batrachion, the first contains pioneer submerged associations which colonise the muddy or silty bottom sediments of disturbed, sheltered waters (drainage ditches, dykes, canals and shallow rivers). Characteristic species are *Ceratophyllum demersum* (rigid hornwort), *Ranunculus circinatus* (fan-leaved water crowfoot), *Hottonia palustris* (water violet) and *Groenlandia densa* (opposite-leaved pondweed).

Associations of the Hydrocharition are restricted in the British Isles to shallow, calcareous, meso- to eutrophic waters in the south and east of England. The character species are *Hydrocharis morsus-ranae* (frog-bit), which occurs locally in shallow mesotrophic waters, and *Stratiotes aloides* (water soldier), an infrequent plant of deeper waters, particularly in eastern counties of England.

Associations of the alliance Callitricho-Batrachion occur in shallow, mesotrophic rivers, streams, ditches and ponds, in which the water table fluctuates considerably during the year. Characteristic species are *Callitriche* spp. (starworts) and members of *Ranunculus* subgenus *Batrachium*, for example, *Ranunculus fluitans* (river water crowfoot) which forms an association with mosses and green algae on the stones, gravel or silt of relatively fast-flowing streams and rivers; *R. circinatus*, which grows in ponds and ditches; and *R. baudotii* (brackish water crowfoot) of brackish, shallow waters near the coast.

6.5.2.3 *Shallow, acid-water communities (Luronio-Potametalia)*
This order contains only one alliance, the Potamion graminei, associations of which colonise more acidic waters than those of either the Magnopotametalia or the Parvopotametalia. They occur in shallow, stagnant or slow- to medium-flowing, non-calcareous, oligotrophic and dystrophic waters, particularly in upland areas. The differential species of the alliance, *Potamogeton polygonifolius* (bog pondweed), occurs with other floating-leaved or submerged species, including *P. gramineus* (various-leaved pondweed), *P. alpinus, Myriophyllum alterniflorum* and *Ranunculus omiophyllus* (round-leaved crowfoot).

6.5.3 *Littoral zone communities of oligotrophic waters (Littorelletea)*

These are amphibian or submerged associations of shallow oligotrophic and mesotrophic lakes, especially in western and northern Britain. The plants form a submerged sward on the gravel and sand near to the shore and also in the deeper waters (to about 4 m) of lakes. In summer when water levels fall the plants of these associations are exposed above the surface. This vegetation is also commonly found in dystrophic waters, in which case the plants are rooted in anaerobic, organic sludge (sapropel). In deeper water they may be replaced by communities of the Luronio-Potametalia.

The class contains one alliance, Littorellion uniflorae, the character species of which is *Littorella uniflora* (shore-weed). Commonly associated plants are *Hypericum elodes* (marsh St John's wort), *Lobelia dortmanna* (water lobelia) and *Subularia aquatica* (awlwort). Additional species include *Isoetes lacustris* and *I. echinospora* (quillworts) – submerged pteridophytes which occur locally in upland oligotrophic to slightly dystrophic lakes.

6.5.4 *Emergent reed and tall sedge vegetation (Phragmitetea)*

This class contains pioneer associations dominated by tall reeds, sedges and grasses which form vegetation mosaics in fens and shallow, open waters (from a few centimetres to more than a metre deep) throughout the British Isles. In river- and lakeside zonations aquatic species of the Potametea and emergent ones of the Phragmitetea occur together in the deeper waters, and in drier situations species more typical of the Parvocaricetea may be present. However, in marginal shallows the emergent species replace the aquatic associations. Characteristic plants are *Phragmites australis*, *Glyceria maxima* (reed sweet-grass), *Iris pseudacorus* (yellow iris), *Sparganium erectum* (branched bur-reed), *Alisma plantago-aquatica* (water plantain), *Rumex hydrolapathum* (water dock), *Lycopus europaeus* (gypsywort), *Berula erecta* (lesser water parsnip), *Rorippa amphibia* (great yellow cress), *Equisetum fluviatile* (water horsetail) and *Myosotis scorpioides* (water forget-me-not).

The large number of associations are classified into three orders: Nasturtio-Glycerietalia (low-growing vegetation of shallow waters and ditches on mineral ground); Phragmitetalia (tall reed-swamp of deeper lakes on substrates often rich in organic matter); and Magnocaricetalia (beds of tall sedges and grasses in shallow, mesotrophic waters).

6.5.4.1 *Communities of dykes, ditches and shallow pools (Nasturtio-Glycerietalia)*

These associations colonise the mineral-rich contact zone between open water and land, where the water table fluctuates considerably throughout the year. In these habitats the water is eutrophic and the diverse vegetation is a mixture of aquatic and emergent plants, including *Nasturtium officinale* (water cress), *N. microphyllum* (narrow-fruited water cress), *Glyceria plicata* (plicate sweet-grass), *G. fluitans* (floating sweet-grass), *Sparganium erectum*, *Myosotis scorpioides* and *Veronica beccabunga* (brooklime). There are two alliances – Glycerio-Sparganion and Apion nodiflori.

Associations of the Glycerio-Sparganion, which are dominated by emergent plants, grow on mineral substrates in shallow waters which usually tend to dry out in summer (ditches and pools). Species of *Glyceria*, *Sparganium* and *Veronica* (including *V. beccabunga*,

6.5 Emergent tall reed vegetation (Phragmitetea): *Phragmites australis*.

V. anagallis-aquatica (blue water speedwell) and *V. catenata* (pink water speedwell)) occur in abundance. In contrast, Apion nodiflori associations, of nutrient-rich, permanently shallow waters, are dominated by a mixture of submerged and emergent species (e.g. *Apium nodiflorum* (fool's water cress), *Scrophularia auriculata* (water figwort), *Nasturtium officinale* and *Callitriche* spp.).

6.5.4.2 Tall reed-swamp (Phragmitetalia)

The tall, emergent vegetation of this order occurs in meso- and eutrophic standing or slow-flowing waters, 0.2 to 2 m deep. The substrates are usually organic, silty muds or sapropel, and the vegetation is often dominated by a single or very few species. Of the two alliances, Phragmition associations occur in stationary or slow-flowing waters and often extend over large areas in and beside lakes, canals and mature rivers; those of the Oenanthion aquaticae are of more localised distribution in shallow, slow-flowing, eutrophic waters overlying calcareous, clayey or sandy substrates.

The associations of the Phragmition are frequently dominated by single species stands of *Phragmites australis*, *Glyceria maxima*, *Typha latifolia* (bulrush), *Schoenoplectus* spp. (club-rush), *Sparganium erectum*, *Iris pseudacorus* or *Scirpus maritimus* (sea club-rush), the last species being common along the silty margins of tidal rivers and brackish standing waters near the coast. Many of the associations of this alliance are fragmentary (i.e. they occupy small areas delimited by very localised habitat conditions), and transient, since sediment deposition or removal continually changes the water depth and trophic status. However, mowing of tall reed beds to provide hay, animal bedding or house thatching prevents successional changes to vegetation of the Parvocaricetea from taking place.

The characteristic species of the Oenanthion aquaticae include *Oenanthe aquatica* (fine-leaved water dropwort), *Sparganium emersum* (unbranched bur-reed), *Sagittaria sagittifolia*, *Rorippa amphibia*, *Butomus umbellatus* (flowering rush) and *Eleocharis palustris* (common spike-rush).

6.5.4.3 Tall grass and sedge beds (Magnocaricetalia)

The single alliance Magnocaricion contains all the associations of tall sedges and grasses of stagnant or very slow-flowing, nutrient-enriched waters overlying organic or mineral substrates. The vegetation of this order usually replaces that of the Phragmitetalia in shallower

a

b

6.6 Vegetation of tall sedge beds: (a) *Carex elata*; (b) *Carex paniculata*.

a   b

6.7   Additional plants of tall sedge beds: (a) *Lythrum salicaria*; (b) *Lysimachia thyrsiflora*.

waters near to the shores of lakes, beside slow-flowing rivers and in fens. There is a considerable range of habitat and species diversity within this alliance, which reflects variations in water depth and nutrient content, drainage, and also the intensity and frequency of mowing; several of the associations are peat-forming.

Characteristic species include *Carex elata* (tufted sedge) and *C. paniculata* (greater tussock sedge), which develop large, compact tussocks on organic substrates at the edge of open water and in fens; *C. acutiformis* (lesser pond sedge) and *C. riparia* (greater pond sedge), which form dense swards on organic-rich, mineral soils by the banks of eutrophic ponds, slow-flowing rivers, canals, ditches and brackish pools, especially in England; *C. aquatilis* (water sedge) which occupies similar habitats to the last two in Wales and Scotland; *C. acuta* (slender tufted sedge) and *C. vesicaria* (bladder sedge) which predominate in the slow-flowing meso- to eutrophic waters of lowland river valleys; the grass *Phalaris*

arundinacea (reed canary-grass), which often forms dense swards along the edges of canals, the flood plains of nutrient-poor rivers and other waterlogged habitats where the summer water table is low; and *Cladium mariscus* (great fen sedge), which colonises nutrient-rich fens and the edges of calcareous lakes. Additional species include *Galium palustre* (common marsh bedstraw), *Lythrum salicaria*, *Lysimachia thyrsiflora* (tufted loosestrife), a local fen plant of northern England and central Scotland, *Poa palustris* (swamp meadow-grass) and *Scutellaria galericulata* (common skull-cap).

6.5.5   *Spring-heads (Montio-Cardaminetea)*

The plant communities of springs are usually dominated by cushions or carpets of bryophytes, within which small vascular plants root. There are two orders: Montio-Cardaminetalia and Cardamino-Cratoneuretalia.

The former contains associations of springs fed by

oligo- to mesotrophic, calcium-poor waters with characteristic species *Montia fontana* (blinks) and the bright-green, acrocarpous moss *Philonotis fontana*. Commonly associated species in the lowland to sub-montane alliance Cardamino-Montion are other small angiosperms such as *Stellaria alsine* (bog stitchwort), *Epilobium palustre* (marsh willowherb), *Cardamine pratensis* (cuckoo flower), *Poa annua* (annual meadow-grass) and *Agrostis stolonifera* (creeping bent), together with, in more shaded situations, *Cardamine amara* (large bitter cress) and *Chrysosplenium oppositifolium* (opposite-leaved golden saxifrage). The numerous bryophytes include: the mosses *Dicranella palustris*, *Brachythecium rivulare*, *Rhizomnium punctatum*, *Calliergon* spp. and *Drepanocladus* spp.; leafy liverworts, in particular *Scapania* and *Solenostoma* spp.; and the thallose species *Pellia epiphylla* and *Conocephalum conicum*. In montane regions this vegetation includes northern and arcticalpine elements and is then referred to the alliance Mniobryo-Epilobion; additional species include *Saxifraga stellaris* (starry saxifrage), *Epilobium alsinifolium* (chickweed willowherb), *E. anagallidifolium* (alpine willowherb) and *Sedum villosum* (hairy stonecrop) (section 9.4.8.1). Spring-head vegetation of the order Montio-Cardaminetalia frequently grades into flush communites of the Caricetalia nigrae (section 6.5.6.1).

The order Cardamino-Cratoneuretalia contains associations of calcareous, meso- and eutrophic spring waters. On contact with the atmosphere at the spring-head, carbon dioxide is lost from the water or is depleted by the photosynthetic activities of the plants growing in the spring, resulting in the production of a hard deposit of calcium carbonate known as 'tuff' or 'tufa'. This is a feature of many springs within this order and the vegetation (particularly mosses) is often thickly encrusted with lime. The most characteristic species of the lowland to sub-montane alliance Cratoneurion commutati is the pleurocarpous moss *Cratoneuron commutatum* which forms a dense yellow-orange mat at the spring-head with several other bryophytes, including *Philonotis fontana*, *P. calcarea*, *Bryum pseudotriquetrum* and *Drepanocladus revolvens*. Vascular species are few, but may include low-growing sedges and grasses, e.g. *Carex panicea* (carnation sedge), *C. flacca* (glaucous sedge) and *Festuca rubra* (red fescue), *Cardamine pratensis*, *Chrysosplenium oppositifolium*, *C. alternifolium* (alternate-leaved golden saxifrage) and the insectivorous *Pinguicula vulgaris* (common butterwort).

In the montane alliance Cratoneureto-Saxfragion aizoidis of calcareous springs, *Saxifraga aizoides* (yellow saxifrage) becomes increasingly dominant (section 9.4.8.1). Vegetation of the order Cardamino-Cratoneuretalia frequently grades into sedge-rich flush communities of the Tofieldietalia (section 6.5.6.2).

6.5.6   *Low sedge communities of transition mires and calcareous fens and flushes (Parvocaricetea)*

Associations of this class occur on oligo-, meso- and eutrophic, nitrogen-poor, mineral substrates where the ground water level is just below or at the surface throughout the year. This vegetation is common in upland areas and is often associated with springs of the Montio-Cardminetea (section 6.5.5). The principal species are low-growing members of the Cyperaceae, Gramineae and Juncaceae, beneath which there is usually a well-developed bryophyte layer.

Characteristic species are *Epilobium palustre*, *Pedicularis palustris* (marsh lousewort), *Stellaria palustris* (marsh stitchwort), *Carex lasiocarpa* (slender sedge), *C. demissa* (common yellow sedge), *C. panicea*, *Potentilla palustris* (marsh cinquefoil), *Hydrocotyle vulgaris* (pennywort) and the bryophytes *Calliergon cordifolium*, *Drepanocladus exannulatus* and *Riccardia pinguis*.

Of the two orders, Caricetalia nigrae contains associations of acidifying transition mire in which the substrates become progressively nutrient-deficient as peat accumulates and the land surface rises above the geogenous water influence; whereas the Tofieldietalia associations colonise calcareous fens and flushes irrigated by nutrient-rich water.

6.5.6.1   *Low sedge vegetation (Caricetalia nigrae)*
The vegetation of this order is 'emersive' in character, i.e. the sedges and other helophytes are rooted in layers of light, fen peat which, when the water table rises, float to the surface, preventing submergence of the above-ground parts of the vegetation. (This is in contrast to the 'immersive' species of the Phragmitetea, which are submerged by rising water levels because they root in deeper, compacted organic and mineral sediments which do not float). Therefore, many species of the Caricetalia nigrae are rooted in peat in which the ground water influence is still effective, whilst acidophilous species (e.g. *Potentilla erecta*, *Vaccinium oxycoccos* and numerous bryophytes) are able to colonise the more acidic, dead plant litter on the peat surface. The associations of the Caricetalia nigrae are intermediate stages in the

6.8 Views of spring heads and flushes: (a) *Philonotis fontana* brown moss flush in Derbyshire; (b) calcareous spring-head tufa mound at Orton Fell, Cumbria.

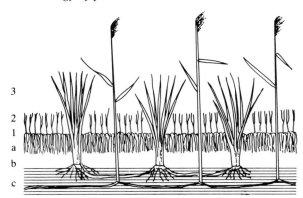

6.9 Rooting differences in a transition mire. The taller, emergent reed grasses and sedges of the Phragmitetalia and Magnocaricetalia are deeply rooted in substrates with a high mineral content. These are being replaced by communities dominated by lower-growing sedges and herbaceous plants of the Parvocaricetea, which are rooted in the upper peat. (Redrawn from Kulczynski, 1949).

a Lightweight, unhumified *Sphagnum-Carex* peat
b Water
c Heavy, humified reed peat
1 *Sphagnum* layer
2 Layer of small perennials (e.g. *Carex limosa*)
3 Layer of large perennials (e.g. *Carex elata, Phragmites australis*)

hydrosere, and in turn, they are replaced (often fairly quickly) by ombrotrophic, *Sphagnum*-dominated associations of the Scheuchzerietea (section 7.3.1) or Oxycocco-Sphagnetea (section 7.3.2).

There is one alliance, Caricion curto-nigrae, the characteristic species of which include *Carex nigra* (common sedge), *C. curta* (white sedge), *C. echinata* (star sedge), *C. lasiocarpa, Molinia caerulea* (purple moor-grass), *Juncus articulatus* (jointed rush), *Ranunculus flammula* (lesser spearwort), *Viola palustris* (marsh violet), *Epilobium palustre, Hydrocotyle vulgaris* and *Potentilla palustris*. The well-developed moss layer includes *Sphagnum* spp. that are tolerant of mesotrophic conditions (including *Sphagnum recurvum, S. teres, S. squarrosum* and *S. palustre*), and several pleurocarpous mosses (principally species of *Calliergon* and *Drepanocladus*). In drier situations small, woody shrubs may also be present, e.g. *Salix repens* (creeping willow), *S. cinerea* (grey willow) and *Myrica gale* (bog myrtle).

6.5.6.2 *Calcareous fens and flushes (Tofieldietalia)*
The species-rich vegetation of calcareous flushes

(usually located below spring-head vegetation of the order Cardamino-Cratoneuretalia (section 6.5.5)) and wet, nutrient-rich fens is in the alliance Eriophorion latifolii of the Tofieldietalia. These habitats are less common in Britain than those associated with the previous order. The vegetation is characterised by the presence of many low-growing members of the Cyperaceae and a large number of other (often rare) flowering plants and bryophytes. Angiosperms include *Eriophorum latifolium* (broad-leaved cotton-grass), *Carex pulicaris* (flea sedge), *C. lepidocarpa* (long-stalked yellow sedge), *Schoenus nigricans* (black bog-rush), *Juncus subnodulosus* (blunt-flowered rush), *Eleocharis quinqueflora* (few-flowered spike-rush), *Pinguicula vulgaris, Epipactis palustris* (marsh helleborine), *Parnassia palustris* (grass of Parnassus), and *Liparis loeselii* (fen orchid). The commoner bryophytes are *Bryum pseudotriquetrum, Calliergon* spp., *Campylium stellatum, Drepanocladus* spp., *Fissidens adianthoides, Pellia fabbroniana* and *Sphagnum contortum* (the only bog moss to tolerate nutrient-rich conditions).

Associations are difficult to delimit owing to the considerable species diversity within the class, the vegetation mosaics that these form and the absence of some of the continental European indicator species from suitable habitats in Britain. However, a number of geographical and environmental variants have been identified. (Wheeler, 1980).

6.5.7 *Alder swamps on peat (Alnetea glutinosae)*

These woodlands are different in physiognomy and species composition from alderwoods on mineral soils, which are discussed in section 5.2.3.1. They develop under the influence of a high water table which, for part or all of the year, rises to the surface producing water-logged conditions, often with open water pools. The vegetation of the Alnetea glutinosae occupies a restricted habitat on organic-rich silts and peats in the transition zone between dry land and ombrotrophic mire, where it replaces associations of the Phragmitetea as peat accumulates but where the ground water influence is maintained. The upper layers of the amorphous peat are neutral to acidic, although the underlying substrates may be rich in nutrients. Many former alder swamps have been drained on account of their fertility and potential for agriculture, yet although reduced in extent these wet woodlands are still widely distributed throughout Britain, especially in valleys, around lakes, swamps

a

b

6.10 Plants of calcareous fens and flushes: (a) *Eriophorum latifolium*; (b) *Carex lepidocarpa*.

and spring-lines. The largest remaining examples are the alder carrs of the Norfolk Broads.

Within the order Alnetalia glutinosae lies the single alliance Alnion glutinosae, which contains all the associations of alder swamp woodland. *Alnus glutinosa* is the canopy dominant; occasional additions to the tree layer include *Betula pendula* and *Fraxinus excelsior*. The shrub layer is usually poorly defined, but may include *Salix cinerea, S. aurita* (eared willow) and *Ribes nigrum* (black currant).

Alder woods on neutral, nutrient-rich substrates often have a species-diverse field layer which includes a mixture of fen and damp woodland species. *Cardamine flexuosa* (wavy bitter cress), *Carex remota, Filipendula ulmaria, Lysimachia nemorum, Valeriana officinalis, Ranunculus repens, Poa trivialis* and *Thelypteris palustris* (marsh fern) are characteristic, as are the bryophytes *Eurhynchium praelongum, Plagiomnium undulatum, Brachythecium rivulare* and *Pellia epiphylla*.

On acidic peat the vegetation is less diverse. *Cirsium palustre, Chrysosplenium oppositifolium, Carex laevigata* (smooth-stalked sedge), *Juncus effusus, Molinia caerulea, Athyrium filix-femina* and *Dryopteris dilatata* are common associates of the field layer, whilst *Sphagnum* spp. are often the dominant bryophytes, for example *S. squarrosum, S. palustre* and *S. recurvum*.

6.5.8 *Alder buckthorn and willow carr (Franguletea)*

As the hydrosere progresses, the upper layers of reed or sedge peat may dry out periodically during the year, so that colonisation of the surface becomes possible by low-growing trees and shrubs. Similar vegetation changes occur on degenerating peat following drainage, disturbance or natural changes in water levels. The class Franguletea includes associations of these wet 'carr forests' on neutral or acid peat. The characteristic species of the single alliance Salicion cinereae are *Frangula alnus* (alder buckthorn), *Salix aurita, S. cinerea, S. aurita × cinerea* and, in northern and western Britain, *Myrica gale*. The moss layer, in which species of *Sphagnum* may dominate, is particularly well developed. Other trees and shrubs are also present, including *Alnus glutinosa, Betula pubescens* and *Viburnum opulus* (guelder rose).

6.5.9 *River bank willow shrub (Salicetea purpureae)*

This class contains pioneer associations of river sides

where the water and underlying substrates are relatively nutrient-rich. These associations are now rare in Britain owing to drainage of flood meadows and river banks and canalisation and related flood control operations. Natural river banks are narrow and irregularly shaped habitats, subjected to high and fluctuating water levels. Their vegetation is very fragmentary and may include associations of other wetland vegetation classes. Characteristic species are *Salix triandra* (almond willow) and *S. viminalis* (common osier), which grow on wet, fairly wide, alluvial flood banks subject to winter flooding, and *S. alba* (white willow) and *S. fragilis* (crack willow) of nitrogen-rich river alluvium in which there is a well-developed gley horizon. These trees grow with a large number of helophytes including *Phalaris arundinacea, Glyceria maxima, Rumex* spp. (dock), *Cardamine amara* and *Iris pseudacorus*. The moss layer is poorly developed.

Chapter 7

Ombrotrophic mires

Ombrotrophic mires occur wherever precipitation exceeds evaporation for most of the year, giving rise to a
permanently high water table. Precipitation is the only
source of water and nutrients to all types of ombrotrophic mire and, consequently, variations in rainfall
and evaporation have a marked effect on their location,
longevity and physical appearance. This is expressed in
the diversity of mire types of the northern hemisphere.
In the Mediterranean region, where for several months
of the year evaporation greatly exceeds precipitation,
mires are almost exclusively rheotrophic (i.e. fens),
whereas in central and northern Europe higher rainfall
and relative humidity promote ombrotrophic peat
development to a depth of several metres.

Ombrotrophic mires are characteristic of the terminal,
acidification stages of the hydrosere, replacing transition
mire vegetation (Chapter 6) as the influence of geogenous water decreases. They originate either in shallow
depressions, wide valleys and watersheds (raised mires)
or in deep, steep-sided depressions (basin mires). In
addition, they may form directly on top of acid, nutrient-
poor, podzolised soils and bare rock, on flat or gently
sloping ground in mountainous or oceanic regions
(blanket mires). At maturity ombrotrophic mires are
inherently unstable and prone to surface desiccation and
erosion.

7.1 The ombrotrophic environment

With the accumulation of peat in mire systems the effect
of geogenous water is removed as the mire surface and
its vegetation become increasingly dependent on a
totally ombrogenous water and nutrient supply. The
substrate becomes more acidic owing to: accumulation
of the acidic by-products from the oxidative metabolism
of organisms living within the confines of the mire; input

of rainfall which is naturally slightly acidic; and exchange of protons for cations in solution (i.e. liberation of H$^+$ ions by cation exchange), which takes place principally by the action of *Sphagnum* spp. (section 7.1.2). Under these conditions of low pH and waterlogging the activity of the soil microflora is limited and the decomposition of organic matter proceeds at a very slow rate (Dickinson, 1983). As a result, layers of undecomposed (unhumified) or partially decomposed (humified) peat accumulate. Some aerobic decomposition takes place at the mire surface and in the upper peat layer where a small amount of oxygen periodically diffuses down, but at depth the peat is saturated with water and anaerobic conditions prevail. Anaerobic decomposition proceeds at a very slow rate, mainly through the activities of obligate and facultative bacteria, yeasts and a few specialised fungi, many of which utilise nitrates, sulphates and phosphates as oxygen sources. Climate can also affect the decomposition rate: during periods of constant high rainfall and high humidity, a peat layer is produced which is unhumified, light brown in colour and contains plant remains which have been virtually unaltered by decay. In contrast, drier climatic periods promote more effective decomposition of the litter layer on the mire surface, so that the rate of peat accumulation is reduced and a dark-brown/black, humified peat is produced in which only small fragments of the peat-forming plants may be visible.

A consequence of the slow rate of decomposition on all ombrotrophic mires is that nutrients are 'locked up' within the peat and are unavailable to living organisms.

Thus, a habitat which is already deficient in all plant mineral nutrients (particularly nitrates and phosphates), as a result of an ombrogenous water supply, is further depleted.

Organic material of any type can hold 100% water by weight, but ombrotrophic peats, and in particular *Sphagnum* peat, can hold as much as 500–600%. They owe this high water-holding capacity to their low bulk densities, high porosities and the highly absorptive, colloidal nature of humified organic matter (Ingram, 1983). Colloidal humus also imparts to peat a high cation exchange capacity when compared to mineral soils.

7.1.1 *Nutrient inputs and losses*

The source of nutrients to ombrotrophic mires is from aerial precipitation (Table 7.1), i.e. the chemical elements dissolved, and particulate material suspended, in rainfall. The inorganic and organic materials brought into a mire ecosystem in this way may originate from a variety of sources (Moore & Bellamy, 1973). In some areas dust raised by vehicles from nearby roads may supplement nutrient inputs; fertiliser applications to agricultural land and dust from quarrying and other earth-disturbing operations may be lifted upwards by wind currents and redeposited on to mire surfaces; and aerosols containing dissolved ions of marine origin may be carried onshore by prevailing winds. Nutrients in the precipitation entering an ombrotrophic mire either become attached to the peat by cation exchange or they are lost from the system by surface run-off waters which

Table 7.1 The chemical composition of rainfall collected at various sites in the UK (values are kg ha^{-1} yr^{-1}).

Sites	Precipitation (cm)	Na	K	Ca	Mg	PO$_4$-P	Inorg-N	NO$_3$-N	NH$_4$-N
Chartley Moss – (Ahmad-Shah, 1984)	92.62	11.75	5.16	6.30	1.92	0.49	7.14	2.23	4.86
Maltby (Dennington & Chadwick, 1978)	63.88	10.37	5.80	17.36	5.30	0.35	7.14	3.03	4.11
Grizedale (Carlisle *et al.*, 1966)	161.69	35.34	2.96	7.30	4.63	0.43	6.28	—	—
Moor House (Gore, 1968)	190.00	32.14	2.27	9.53	4.48	0.27	6.89	—	—
Kent (Madgwick & Ovington, 1959)	84.00	19.30	2.80	10.70	<4.20	<0.40	—	—	—
Abbots Moss (Allen *et al.*, 1968)	95.00	14.0	5.40	14.0	2.9	0.80	—	—	—
Silpho (Allen *et al.*, 1968)	70.00	36.0	4.80	9.8	4.5	0.80	—	—	—
Merlewood (Allen *et al.*, 1968)	150.00	22.0	3.70	12.0	3.6	1.00	—	—	—
Kerlock (Allen *et al.*, 1968)	95.00	28.0	3.90	6.7	3.6	0.20	—	—	—
Craigeazle (Boatman *et al.*, 1975)	206.00	34.5	3.30	5.2	4.7	—	—	—	—

Table 7.2 Input of chemical elements to Chartley Moss in the precipitation under different tree canopies and in the open (values are kg ha^{-1} yr^{-1}).

	Na	K	Ca	Mg
Tree litter, dust and faeces				
Mixed woodland	1.01±0.22	10.17±1.60	28.89±8.28	6.10±1.24
Fen woodland	1.04±0.14	10.91±1.69	43.97±5.30	12.88±1.34
Pine woodland	1.49±0.22	12.29±2.49	45.51±8.95	8.28±1.73
Precipitation				
Sphagnum lawn	16.03±1.85	9.22±4.88	7.77±1.95	2.44±1.05
Mixed woodland	18.67±3.20	8.33±1.43	12.82±1.66	4.74±0.87
Fen woodland	20.94±2.92	17.42±3.28	24.05±3.44	11.20±1.70
Pine woodland	23.36±5.02	21.57±4.74	24.82±3.92	7.06±1.19
Total aerial inputs				
Sphagnum lawn	16.03	9.22	7.77	2.44
Mixed woodland	19.68	18.50	41.71	8.54
Fen woodland	21.98	28.33	68.02	24.08
Pine woodland	24.85	33.86	70.33	15.34
Ratio of nutrients under tree canopies to that in Sphagnum lawn				
Mixed woodland	1.2	2.0	5.4	3.5
Fen woodland	1.4	3.1	8.8	9.9
Pine woodland	1.6	3.7	9.1	6.3

flow into marginal streams and drainage ditches.

The input of chemical elements to the surface of forested ombrotrophic mires is greatly enhanced by the presence of a tree canopy from which chemical elements contained in, or particulate material impacted upon, the leaves are removed by rainfall and carried to the mire surface (Rieley *et al.*, 1984). Of three tree canopies studied at Chartley Moss NNR in Staffordshire (*Pinus sylvestris*, *Alnus glutinosa/Salix* spp. and *Betula pubescens/Sorbus aucuparia*), all enhanced the nutrient input to the mire surface, compared with an open area of *Sphagnum* lawn, but the greatest enrichment of all elements, with the exception of magnesium, occurred under a dense canopy of *Pinus sylvestris*. Part of this nutrient input is derived from the droppings of *Corvus frugilegus* (rook), a large colony of which roost over-winter in the pine trees on this nature reserve (Table 7.2).

The nitrogen input to ombrotrophic mires may be enhanced by the activities of free-living nitrogen-fixing organisms (bacteria and blue-green algae (Cyano-bacteria)). Although these are most abundant in soils with a neutral or slightly alkaline pH, they have been recorded in small numbers from ombrotrophic mires, usually where the pH is greater than 4. Cyanobacteria may be abundant in the open water pools of acid mires; these nitrifying algae are also known to form more intimate endophytic associations with *Sphagnum* spp.

and have been recorded from within the hyaline cells (see below) of these mosses (Brown, 1982). However, the importance of this fixation to the nitrogen nutrition of *Sphagnum* spp. and higher plants or other mosses 'rooting' amongst the *Sphagnum* has yet to be clearly established, as has the exact role of this fixation in the nitrogen economy of these mires.

Certain processes may result in the loss of nutrients from ombrotrophic mires (Table 7.3). For example, burning of vegetation accelerates the movement of chemical elements since a large proportion of the plant nutrients contained in standing vegetation are released in ash or smoke (Allen, 1964; Gimingham, 1972). When *Calluna vulgaris* is set on fire, over half the carbon, nitrogen and sulphur in the vegetation are driven off in the smoke and, in addition, several mineral elements, especially potassium, are readily leached from the re-sidual ash, although losses may be reduced as a result of adsorption on to *Sphagnum* spp. by cation exchange (section 7.1.2). Peat erosion also results in another major loss of nutrients from mires (section 7.1.6) and smaller nutrient losses occur as a result of sheep grazing (Tallis, 1983; Crisp, 1966).

Plants of ombrotrophic mires, therefore, have two major handicaps to colonisation and survival – waterlogged (i.e. poorly aerated), acidic 'soils' and extremely low availability of mineral nutrient elements. The plants which thrive best under these conditions are

Table 7.3 Water and nutrient balance sheet for moorland in England (after Crisp, 1966).

	Water (thousands of m³)	Na (kg year⁻¹)	K (kg year⁻¹)	Ca (kg year⁻¹)	P (kg year⁻¹)	N (kg year⁻¹)
Stream water output	1368	3755	744	4461	33	244
Evaporation	403	—	—	—	—	—
Peat erosion	—	23	171	401	37	1214
Drift of fauna in stream	—	0.004	0.011	0.003	0.010	0.118
Drift of fauna on stream	—	0.11	0.38	0.07	0.43	4.6
Sale of sheep and wool	—	0.16	0.44	1.58	0.98	4.4
Total output	1771	3778	916	4864	71	1467
Input in precipitation	1771	2120	255	745	38–57	681
Difference = Net loss for catchment	—	1658	661	4119	14–33	786
Net loss ha⁻¹	—	20.01	7.97	49.68	0.17–0.40	9.48

not vascular species but members of the genus *Sphagnum* (bog mosses), which are major peat-forming species well adapted to the ombrogenous environment.

7.1.2 Ecology of Sphagnum

These mosses have two features which make them particularly well equipped to grow on ombrotrophic mires – a high water-holding ability and a high cation exchange capacity. In common with other bryophytes *Sphagnum* spp. have no internal water conducting tissues. Water is absorbed directly via a network of capillary spaces in the moss plant. This upward movement of water in and between the closely packed stems and branches of individual plants enables certain species to grow above the mire water table and form large hummocks which may exceed 0.5 m in height. Sphagna also have a high water storage capacity as a result of their peculiar leaf structure: small chlorophyllose (photosynthetic) cells alternate with larger hyaline (colourless) cells, which have pores in their cell walls through which water can move. At maturity the hyaline cells are dead and are capable of absorbing and retaining considerable quantities of water. This high water-holding capacity of living *Sphagnum* plants growing on the mire surface, added to the capillary rise of water within the peat, ensures that the water table is maintained at or near the mire surface throughout the year. (Smith, 1982).

Sphagnum species are able to flourish particularly well in ombrogenous situations as a result of their relatively high cation exchange capacity (CEC), i.e. their ability to take up cations in solution by exchanging protons (H⁺) held at exchange sites in the plant. This is in common with other bryophyte species. The exchange sites in *Sphagnum* appear to be carboxyl groups (COO⁻) located in the cell wall on polymers of uronic acids (Clymo, 1967). As rainwater passes through *Sphagnum* it becomes more acidic as a result of cation exchange until all the H⁺ ions originally present as COOH have been liberated into solution. The ability of *Sphagnum* to maintain a high cation exchange capacity and hence continue to lower the pH of their surroundings probably requires continual growth of the moss plants, i.e. production of new exchange sites. Uronic acids can constitute up to a third of the dry mass of *Sphagnum* plants. It has been shown (Clymo, 1963 and Clymo & Hayward 1982) that there is a correlation between the uronic acid content (i.e. cation exchange capacity) of *Sphagnum* species and the height of their optimum habitat above the mire water table. For example, *Sphagnum cuspidatum*, which grows in pools, contains about 10% uronic acids and has a lower CEC per unit weight than *Sphagnum capillifolium*, which inhabits drier hummocks and contains about 30% uronic acids. A secondary effect of their high CEC is that *Sphagnum* hummocks usually have a lower pH than waterlogged pools.

7.1.3 Conservation of nutrients by vascular plants

Nutrient conservation in plants is of great importance under conditions of low nutrient supply. Processes which contribute to the retention of nutrients in the plants growing on peat include: return of nutrients to perennating organs at the end of the growing season and the relocation of elements to regions of new growth the following year; penetration by roots down to deep peat zones and the transport of nutrients from these to the surface. In *Eriophorum vaginatum* (hare's tail cotton-grass), for example, phosphorus and potassium are transferred from dying leaves to the rhizome, thus conserving nutrients which are in short supply and, in addition, the roots of this species may extend downwards for more than a metre (Goodman & Perkins, 1959). Other processes increase nutrient availability to mire plants: mycorrhizal associations (a large number of the vascular species of ombrotrophic mires, including most members of the Ericaceae, some grasses, *Pinus sylvestris* and *Betula pubescens* are known to be infected with mycorrhizal fungi which facilitate the uptake of nutrients, especially phosphorus, from organic sources); nitrogen fixation (e.g. *Myrica gale* (bog myrtle), a small shrub of wet, ombrotrophic mires and fens, has root nodules inhabited by nitrogen-fixing actinomycetes); and insectivorous habit (e.g. *Drosera* spp. (sundews) may supplement their nutrient supply with small insects caught on sticky glands on their leaves).

7.1.4 *Waterlogging*

Most of the vascular species that inhabit ombrotrophic mires have special adaptations, particularly to their roots, which enable them to tolerate anoxia and the phytotoxic levels of certain compounds and ions that are present in anaerobic, acidic substrates. (Adaptations to water excess in soils are discussed in section 1.1.3.2.) Monocotyledonous species (grasses and sedges), which form a major part of the vegetation of most mires, exhibit seasonal dimorphism in their roots. The root system of these plants is adventitious and dies back at the end of each growing season prior to the most severely waterlogged winter period. When water levels drop below the peat surface in the spring, new roots develop in the surface oxygenated layer of peat. These adventitious roots are unlignified, shallow-rooted and spreading, features which enable oxygen diffusion into the plants and ethanol dissipation from them to take place. Many ericaceous shrubs are also shallow-rooted. (Moore & Bellamy, 1973).

Mire dicotyledonous species (particularly members of the Ericaceae) frequently have xeromorphic, sclerophyllous leaves and stems in addition to root modifications; and most monocotyledonous plants found on mires are xeromorphic, e.g. *Eriophorum*, *Rhynchospora* and *Trichophorum* spp. The exact reasons for these morphological features are unclear: the reduction of transpirational water losses through leaves and stems

may reduce the root uptake of toxic ions in solution, thus avoiding or postponing the lethal consequences of root poisoning; or the xerophytic habit may represent the most economical means of producing photosynthetic tissues in an oligotrophic environment. (Gore, 1983).

7.1.5 *Peat formation processes*

The manner in which ombrotrophic peat accumulates, often to considerable depth above the geogenous influence has long been a subject of interest. One theory proposes an alternation, on the mire surface, of waterlogged depressions and relatively dry hummocks ('lenticular regeneration') (Kulczynski, 1949) (Fig. 7.1). This assumes that the species of Sphagna which grow submerged in wet pools and hollows (e.g. *Sphagnum cuspidatum* and *S. auriculatum* var. *inundatum*) grow faster than either the species which colonise the water surface to form more or less homogeneous lawns (e.g. *S. magellanicum* and *S. subnitens*) or those that develop hummocks emergent above the mire water table (e.g. *S. capillifolium*, *S. fuscum*, *S. imbricatum* and *S. papillosum*). It is envisaged that the hollows gradually infill and develop into low hummocks, whilst the older, larger hummocks degenerate following colonisation by species tolerant of drier conditions (e.g. other bryophytes, lichens, cotton-grasses and ericaceous shrubs). Eventually, the regressing hummocks become waterlogged and form the sites of new hollows, partly as a result of their collapse and partly because the water table continues to rise upwards by capillarity in the surrounding, actively growing bog mosses. This cyclic process, if it were to operate for long enough, would

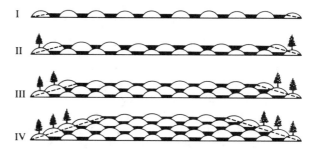

7.1 Proposed scheme for the lenticular regeneration of *Sphagnum* peat. The black areas denote regeneration hollows (e.g. *S. cuspidatum*), white indicates regeneration hummocks (e.g. *S. fuscum*). The broken line represents the water table in the peat, rising as regeneration proceeds. (Redrawn from Kulczynski, 1949).

lead to the accumulation of a layer of ombrogenous peat up to several metres deep.

Unfortunately, evidence to support this theory is rather inconclusive and, although some stratigraphical information suggests that it could explain the sequence of raised mire development in parts of central Europe, southern Scandinavia and lowland Britain, all of which are regions of relatively low rainfall, this -has not been confirmed by studies of peat cores from ombrotrophic mires throughout Europe. In regions of very high rainfall, e.g. in western Ireland and north-west Scotland, for example, the peat on some blanket mires has apparently increased in height uniformly over the entire mire surface. The pools and depressions have remained more or less in the same position throughout their development and, therefore, have not alternated in the manner suggested for raised mires. (Boatman, 1983).

7.1.6   *Peat erosion*

The upward growth of peat on ombrotrophic mires is limited by the amount of water supplied to the mire surface from rainfall, and the rate of capillary water rise within the peat mass. If these are insufficient to maintain the active growth of Sphagna then the rate of peat accumulation slows down and the surface of the mire dries out and may become susceptible to erosion.

Long-term climatic changes also affect the rate at which ombrotrophic peats accumulate. The climate of the British Isles is now considerably drier than it was in the Atlantic period (8000–5000 BP) or the early sub-Atlantic period (2800–2000 BP) when the most rapid accumulation of peat occurred and, as a consequence of this, slow but consistent erosion may have been proceeding in parts of England and Wales throughout at least the last 1000 years. However, the onset of rapid peat erosion is a relatively recent phenomenon – perhaps within only the last 200 years. (Bower, 1962).

Most raised mires are contained within a definite catchment and regression of the peat is a marginal feature which is pronounced only if the mire has been cut over, drained and/or afforested. Blanket mires, on the other hand, are very vulnerable to erosion, particularly those at the southern and eastern limits of blanket mire distribution in Great Britain, where the present effective rainfall is insufficiently high to promote peat development. In many areas of blanket peat the annual rainfall is now less than the minimum considered necessary to maintain a uniform cover of *Sphagnum* spp. (less than 1,500 mm per annum (Pearsall, 1971)); linked with

a low relative humidity, this can result in long periods during the year when the peat surface dries out (Fig. 7.2). There have been, however, many other factors implicated in peat erosion in addition to climatic change. These include trampling by man and his domestic animals, pollution, grazing and burning and it is likely that several factors are responsible for the onset of erosion. For example, on the moorlands of the southern Pennines many *Sphagnum* spp. have disappeared directly as a result of aerial pollution from the surrounding conurbations, a change that has taken place during the 150 years since the start of the Industrial Revolution (Ferguson *et al.*, 1978).

When a peat surface dries out, oxidation and shrinkage occurs, cracks appear on the surface and these fill with rainwater. Expansion of this water as it freezes causes the fissures to increase in size, exposing more peat to the atmosphere during subsequent periods of low rainfall. Exposed, amorphous (structureless) peat, which is no longer able to absorb water, may then be carried away in run-off water or blown away by strong winds. Blanket mires which reach this advanced stage of erosion are unable to recover and their eventual destruction is ensured. (Tallis, 1983) (Fig. 7.2).

7.1.7   *Historical information contained in peat deposits*

Paleoecology involves the reconstruction of past regional and local floras through the identification of fossil plant remains. These include not only the petrified impressions and casts of ancient land plants, but also the virtually unaltered remains (macro- and microfossils) found in unconsolidated lake sediments and mire peats, which shed light on more recent (i.e. post-glacial) vegetation changes. These latter fossils are preserved in waterlogged, anaerobic sediments in which organic decomposition is minimal.

In peat deposits, the plant remains are incorporated into a largely organic matrix, whereas lake sediments contain both organic and finely grained mineral matter. Commonly, deposition has occurred sequentially, so that investigation of the fossils found in a vertical sample of a sediment allows the order of past regional and local vegetational changes to be deduced.

Samples are collected with some type of coring device, such as a peat borer, by which consecutive cores can be removed from the peat or lake deposits. The plant remains can be conveniently divided into micro- and macrofossils: the former are microscopic pollen grains and spores, whilst the latter include larger fragments of

7.2   Peat erosion: (a) aerial view of blanket peat erosion in Glen Torridon, Wester Ross; the anastomosing channels drain into the River Torridon; (b) peat haggs on Crowden Head, Peak District National Park.

stems, leaves, roots, seeds and fruits. The investigation of microfossils is known as 'palynology' or pollen analysis (Moore & Webb, 1978). The resistant nature and distinctive shape and pattern of the outer coat (exine) of both pollen grains and the spores of lower plants (bryophytes and pteridophytes) ensure that they are well preserved and that following physical and chemical treatments, they can be accurately identified and counted (Fig. 7.3). Within certain limits (the most important being that pollen grains are wind-borne and so may have travelled over a great distance), the data obtained from pollen analysis, which are usually represented in pollen diagrams, can be used to illustrate past changes in the vegetation of the region surrounding the site, both quantitatively and qualitatively (Fig. 5.1). The age of polleniferous deposits can be dated fairly accurately by comparison with other documented sequences and by radioactive carbon dating. Palynology is also increasingly used to trace the history and distribution of individual species (section 9.3).

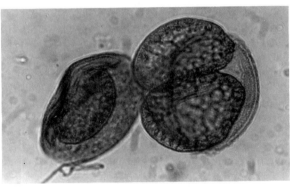

a

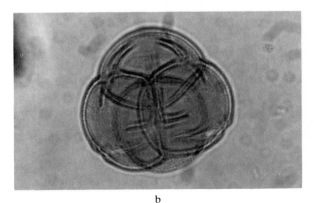

b

7.3 Pollen grains commonly found in peat deposits: (a) *Pinus sylvestris*; (b) Ericaceae.

The examination of macrofossils can only provide information on very localised transformations in plant assemblages, since, unlike microfossils, their dispersal is limited. Macrofossil data, together with information on the structure of the sediment (i.e. texture, colour, degree of humification and wetness) can be used to illustrate the successional development of the deposit and the plant communities that it supported at different times in the past. These 'stratigraphical' studies can provide valuable information on hydroseral successions and mire development, as well as aiding mire classification (Fig. 7.4).

7.2 Ombrotrophic mire types and their distribution in Britain

Throughout Europe, there is a general zonation of ombrotrophic mires from south to north and from east to west, determined by variations in climate. However, within the British Isles, the three major ombrotrophic mire types often occur in close proximity to each other owing to variations in rainfall and topography over relatively short distances, and it may be difficult without detailed stratigraphical information to distinguish between raised and blanket mires in upland and high rainfall districts, especially since the surface vegetation of both is similar. (Moore & Bellamy, 1973).

7.2.1 *Raised mires*

Ombrotrophic raised mires are scattered throughout Britain, but occur most frequently in inland situations at low altitudes (less than 300 m) and in coastal regions where rainfall is moderate but humidity is high. They are absent from the south-east of England where rainfall is too low and evaporation is high. Although most have suffered considerably from drainage, peat-cutting, burning, pollution and afforestation, some excellent examples can still be found in west Wales, Cumbria and Galloway (around the Solway Firth), in the west Midlands plain, north-east England and central and north-east Scotland. (Taylor, 1983).

Raised mires are characteristically 'domed' and the surface of the elevated, convex, central cupola of peat may be many metres above the mineral ground of the mire basin (Fig. 7.5). The height of this cupola is related to the amount of annual rainfall and raised mires have a greater convexity in regions of high rainfall. Although ombrotrophic peat is low in nutrient el-

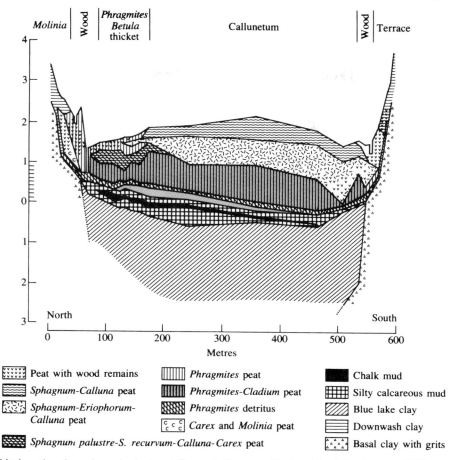

7.4 Stratigraphical section through a raised mire at Rosgoch Common, Radnorshire (after Bartley, 1960).

ements, the edge zone of a raised mire (lagg, from the Swedish) receives drainage water from the surrounding mineral ground, which enhances the ionic content of the marginal waters. The lagg water table fluctuates vertically throughout the year and, after heavy rain, water circulates around the perimeter and flows out of the catchment in one or more drainage streams. The sloping, elevated edge of the mire near to the lagg (the rand – Swedish), is much drier than the central cupola and may be wooded with trees of *Pinus sylvestris* and *Betula pubescens*. Here the water table may be as much as half a metre or more below the surface since the gradient of this marginal zone encourages water movement from rand to lagg, and dehydration is further promoted by transpiration of the trees. The central, most elevated, and wettest, region of the mire, where the water table is permanently at or near the surface and which is dominated by different species of *Sphagnum*, usually consists of a complex of hummocks and waterlogged hollows (Fig. 7.6). Small open-water pools are common, but large pools, if present, are a secondary development following erosion or degeneration of the surface peat.

The central cupola of some raised mires is distinctly 'patterned', with elongated pools developed along lines of stress on the peat surface and orientated at right angles to the direction of slope. On raised mires which originated in almost symmetrical, shallow lakes or watershed catchments, these pools are concentrically positioned around the central cupola. These 'concentric raised mires' often extend over large areas and may have a peat depth in excess of 6 m. On the raised mires of broad valleys in upland or northern regions where the peat has formed over small, shallow (usually

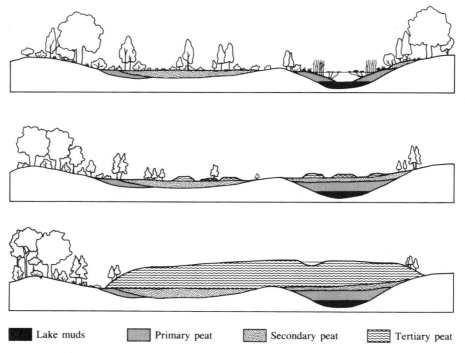

7.5   Schematic representation of the development of an ombrotrophic raised mire from a eutrophic lake and adjacent fen woodland (redrawn from Ellenberg, 1963).

7.6   Ombrotrophic raised mire: Borth Bog, west Wales.

7.7 Patterned ombrotrophic mire: Silver Flowe, Galloway, south-west Scotland.

glaciated) depressions on gently sloping ground, the surface bog pools are eccentrically arranged. The best examples of 'eccentric raised mires' are in Scandinavia (especially Finland), although Claish Moss and Silver Flowe in Scotland are of this type (Fig. 7.7).

7.2.2 Basin mires

The term 'basin mire' is confusing. In its strict sense it should only be applied to Schwingmoore (German – swinging bogs) formed in steep-sided, water-filled basins, over which a mat of vegetation extends forming a floating raft of peat that encloses a large sub-surface reservoir. At maturity, these 'floating mires' often support forest on the drier, firmer, marginal peat, but the unstable, central regions, which are strongly influenced by vertical fluctuations in the level of the sub-surface water, are treeless. In this floating section the water table is always maintained at or near the surface of the mire owing to synchronous oscillations of the peat raft.

However, the term 'basin mire' is frequently used in a very general way to describe any ombrotrophic mire which occupies a steep-sided (often quite small) basin, or land-locked valley, irrespective of the origin and development of the mire, or even if a true floating raft of peat is present or not. For example, Chartley Moss, Staffordshire, and Wybunbury Moss, Cheshire, which

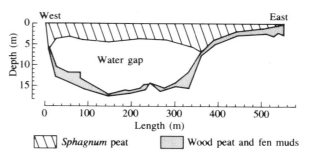

7.8 Basin profile of Wybunbury Moss, Cheshire (modified from Green & Pearson, 1977).

are often referred to as the best examples of floating mires in Great Britain, are not Schwingmoore at all but are 'collapsed raised mires' which occupy subsidence hollows (Figs 7.8 and 7.9). Initially, these developed as raised mires within shallow, glacial depressions. However, at a later stage subsidence occurred following solution of underground saliferous strata. The peat collapsed and split, with the upper layer of lighter *Sphagnum* peat separating from the lower, denser (heavier) fen peat. The former floated on the surface of the water in the newly formed deep basin, enclosing a sub-surface reservoir of water, while the latter sank to the bottom. (Rieley *et al*, 1984).

Another unusual ombrotrophic basin mire which has

7.9 Chartley Moss, Staffordshire – a collapsed raised mire, the best British example of a basin mire with a floating raft of *Sphagnum* peat.

been found in the English Lake District (but which may be of more widespread occurrence in Britain), is the 'sediment' or 'erosion' mire. Mires of this type have developed in a few steep-sided, upland tarns which have become filled with allochthonous peat sediments eroded from blanket mires at higher elevations. These organic deposits can be very deep (up to 16 m) and consist of a thick layer of almost structureless, compact, well-humified peat, overlying mineral lake sediments or rock. On the surface of this sediment peat is a shallow layer

of poorly humified, ombrotrophic peat, in which there are abundant *Sphagnum* remains. This has been formed by the acidophilous plant communities which colonised the mire surface after the open water of the tarn disappeared. Deep peat deposits in upland parts of the Lake District are unusual because the oligotrophic nature of the lake waters does not promote high plant productivity or peat accumulation.

7.2.3   *Blanket mires*

Blanket mires are restricted to those regions of the north and west of the British Isles where either rainfall is very high (greater than 1500 mm), or evaporation is low (precipitation: evaporation ratio greater than 1). These mires have developed in shallow depressions and on flat and gently sloping ground at altitudes up to 1000 m and, although they form the characteristic 'capping' to elevated mountain shoulders, they also occur at sea level in western Ireland and north-west Scotland.

Blanket mires are the result of 'paludification', whereby peat formation is initiated directly on waterlogged, acid, mineral substrates in which drainage is impeded (e.g. podzols with an iron pan) or on impervious, acidic rocks (Figs 7.10 and 7.11). This is in contrast to the process of terrestrialisation by which raised mires form in shallow lakes and basins. Many blanket mires on sloping ground in western and northern regions of the British Isles have a distinct pattern of elongated pools on their surface similar to that already described for raised mires. This 'patterning' may develop under the influence of gravity as the peat moves slowly downhill, causing the surface to split. The best known examples are in the 'Flow' country of Caithness and Sutherland in northern Scotland.

7.3   Vegetation of ombrotrophic mires

Despite the range of ombrotrophic mire types in the British Isles, the floristic composition is remarkably similar throughout and the plant associations are contained within only two classes – Scheuchzerietea and Oxycocco-Sphagnetea. The species of the former are

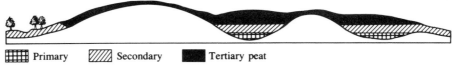

⊞ Primary   ▨ Secondary   ■ Tertiary peat

7.10   Diagrammatic representation of blanket mire formation in north-western Scotland.

7.11 Part of the extensive blanket peatland of the Caithness Flows in northern Scotland.

characteristic of open-water pools and waterlogged depressions, whilst those of the latter colonise hummocks and elevated peat in which the water table is below the surface throughout the year. The vegetation of both classes grows in complex mosaics (usually in close proximity to each other), which reflect variations in substrate hydrology and nutrient content, site altitude and proximity to the sea (oceanicity).

7.3.1 Vegetation of waterlogged pools and hollows (Scheuchzerietea)

This is the vegetation of the central hollows, lagg and secondary erosion pools of ombrotrophic mires. The characteristic species of the class, the single order Scheuchzerietalia palustris and the one alliance Rhynchosporion albae is *Scheuchzeria palustris* (Rannoch-rush), a species which used to be more widely distributed throughout the British Isles than it is at the present day. As a result of drainage and destruction of suitable habitats it now occurs only in a few locations on Rannoch Moor, Scotland. The principal vascular plants of this class are *Carex limosa* (bog sedge), *Rhynchospora alba* (white beak sedge), the insectivorous *Drosera anglica* (great sundew) and *D. intermedia* (oblong-leaved sundew), the bog mosses *Sphagnum cuspidatum*, *S. recurvum* and *S. pulchrum* (mainly confined to western Britain) and the liverwort *Cladopodiella fluitans*.

The shallow pools of the hummock-hollow complex support a continuous carpet of *Sphagnum cuspidatum* in which *S. auriculatum* var. *inundatum* occurs frequently. *Eriophorum angustifolium* (common cottongrass) and *Rhynchospora alba* are common, and the latter is abundant in the narrow marginal zone between the edges of the pools and the bases of the taller *Sphagnum* hummocks. In north-west Scotland and western Ireland *Carex limosa* is also present. The deep pools formed as a result of surface erosion and the elongated pools of patterned mires support a sparse vegetation, mainly of *Menyanthes trifoliata* (bogbean), *Sphagnum cuspidatum* and *Utricularia* spp. (bladderwort) on their muddy bottoms. These small dystrophic pools often contain dense growths of the algae *Batrachospermum* spp., *Rhizoclonium* spp. and *Zygogonium ericitorum* which, following drying out or drainage, form a crust on the peat surface. *Rhynchospora fusca* (brown beak-sedge) and *Hammarbya paludosa* (bog orchid) are of infrequent occurrence in shallow bog pools, particularly in the New Forest, the central and western Scottish Highlands and the Outer Hebrides. The rarest plant of this pool habitat in the British Isles is the Atlantic species *Eriocaulon aquaticum* (pipewort), which only grows in a few blanket mires in western Ireland and the Inner Hebrides of Scotland.

A species-poor variant of this vegetation develops on the central oligotrophic bog pools and soaks of basin mires where the water table oscillates vertically in response to seasonal variations in rainfall. The dominant plant is *Sphagnum recurvum* on the surface of which *Vaccinium oxycoccos* (cranberry) grows and through which the leaves of *Eriophorum angustifolium* appear in profusion. *Eriophorum angustifolium* and *Sphagnum recurvum* also colonise the open-water pools or bare peat which form after peat cutting or erosion.

7.3.2 *Peat-forming vegetation of ombrotrophic mires and wet heaths (Oxycocco-Sphagnetea)*

The vegetation of this class, dominated by ericoid shrubs, members of the Cyperaceae and an abundance of *Sphagnum* spp., is widely distributed throughout northern Europe. The characteristic species are *Drosera rotundifolia* (round-leaved sundew), *Trichophorum cespitosum* (deer-grass), *Eriophorum vaginatum*, *Narthecium ossifragum* (bog asphodel), *Erica tetralix* (cross-leaved heath), *Andromeda polifolia* (bog rosemary) and the bryophytes *Aulacomnium palustre*, *Sphagnum tenellum*, *S. capillifolium*, *S. papillosum*, *Calypogeia trichomanis* and *Lepidozia setacea*. Of the two orders, Ericetalia tetralicis and Sphagnetalia magellanici, associations of the former occur on shallow, nutrient-poor ombrotrophic peats; the latter order is mainly composed of peat-forming associations of the deeper, raised peat of hummocks and lawns on blanket and raised mires (Moore, 1968).

7.3.2.1 *Wet-heath vegetation (Ericetalia tetralicis)*
The character species of this order and the alliance Ericion tetralicis are *Erica tetralix*, *Trichophorum cespitosum*, *Juncus squarrosus* (heath rush), the bog mosses *Sphagnum compactum* and *S. tenellum*, the leafy liverwort *Gymnocolea inflata*, and the green alga *Zygogonium ericitorum*. The vegetation is relatively species-poor but contains many bryophytes.

In high rainfall areas near Atlantic coasts wet heaths dominated by *Trichophorum cespitosum* and *Calluna vulgaris* are common on gentle slopes which are too well drained to support the vegetation of the Erico-Sphagnion which grows on wetter peats. The most frequently associated species are *Erica tetralix* and *Potentilla erecta*. There are a number of sub-types of this vegetation which reflect local differences in climate and physiography: in depressions at high altitude *Sphagnum palustre*, *S. papillosum* and *S. capillifolium* are common; following disturbance and desiccation the *Sphagnum*

cover may be partly replaced by hummocks of *Racomitrium lanuginosum*; and bare peat surfaces are often colonised by lichens (e.g. *Icmadophila ericitorum*) and liverworts (e.g. *Lepidozia setacea*, *Mylia anomala* and *Odontoschisma denudatum*).

In the Scottish Highlands, on sloping ground with shallow blanket peat, *Molinia caerulea* occurs in abundance together with *Erica tetralix*, *Narthecium ossifragum*, *Potentilla erecta* and *Trichophorum cespitosum*. Within this vegetation several species associated with lowland rheotrophic mires are common. These plants, which indicate an increased nutrient enrichment from edaphic or aerial sources, include *Carex panicea* (carnation sedge), *Pinguicula lusitanica* (pale butterwort) and *Schoenus nigricans* (black bog-rush).

7.3.2.2 *Hummock-forming Sphagnum-dominated vegetation (Sphagnetalia magellanici)*
This order contains the main peat-forming, *Sphagnum*-dominated associations of ombrotrophic mires on peat more than 1.5 m deep. There are two alliances: Erico-Sphagnion, which comprises *Sphagnum*-dominated hummock and lawn vegetation of raised mires near Atlantic seaboards (section 7.2.1), and blanket mire vegetation of upland regions of Great Britain and coastal areas of north-west Scotland and western Ireland (section 7.2.3); and Sphagnion fusci, which is characteristic of the terminal stage of the raised mire succession dominated by hummocks and lawns of *Sphagnum fuscum*. This latter alliance is poorly represented in the British Isles and achieves its optimum expression in central and eastern Europe.

The associations of the Erico-Sphagnion constitute the central 'flat' and low hummock communities of mires in regions where the snow cover is limited to a few weeks of the year and from which *Sphagnum fuscum* is either absent, or occurs only as isolated, low hummocks. The character species are *Erica tetralix*, *Narthecium ossifragum*, *Myrica gale*, *Drosera intermedia*, *Sphagnum magellanicum*, *S. papillosum*, *S. subnitens*, *S. imbricatum* and *Odontoschisma sphagni*. Throughout the geographical range of this alliance *Andromeda polifolia* and *Vaccinium oxycoccos* are both absent. Associations of the Erico-Sphagnion replace vegetation of the Scheuchzerietea as wet pools and hollows become filled in as a result of peat accumulation. In addition they are characteristic of deep humified peats on level or gently sloping ground where the peat has

7.12 Plants of the waterlogged hollows of ombrotrophic mires: (a) *Scheuchzeria palustris*; (b) *Carex limosa*; (c) *Rhynchospora alba*; (d) *Menyanthes trifoliata*

been cut in the past or drained to provide rough grazing. This type of land is now favoured for forestry.

Two geographical variants of this alliance have been distinguished in the British Isles (Moore, 1968):

1. An extreme 'Atlantic' type of western Ireland and north-west Scotland in which *Sphagnum imbricatum*, *Pleurozia purpurea* and *Erica tetralix* are common. *Eriophorum vaginatum* becomes rarer in the far west

7.13 Ombrotrophic mire species: (a) *Narthecium ossifragum*; (b) *Eriophorum vaginatum*; (c) *Aulacomnium palustre*; (d) *Andromeda polifolia*.

of the British Isles and near the coast of western Ireland and north-west Scotland it is replaced by *Schoenus nigricans*.

2. A 'sub-Atlantic' type found in England and Wales in which *Narthecium ossifragum* is common and *Sphagnum recurvum* is present, often in considerable abundance. Although *Sphagnum imbricatum* is no longer present in the surface vegetation of these mires, fossilised remains of this plant are usually present in the sub-surface peat. The disappearance of this last species has been related to climatic change (i.e. decreased rainfall and oceanicity), which has reduced the input of nutrient elements to ombrotrophic mires, and also to aerial pollution from industrial sources which may have introduced toxic elements into the mire environment (Green, 1968).

In the extensive hill-lands of western and northern Britain, for example the Pennines, the Lake District, the Cheviots, north Wales and the southern uplands of Scotland, the dominant vegetation of the Erico-Sphagnion has certain floristic affinities with associations of the continental-boreal Sphagnion fusci alliance. *Empetrum nigrum*, *Eriophorum vaginatum*, *Calluna vulgaris* and the bryophytes *Pleurozium schreberi*, *Hylocomium splendens* and *Sphagnum capillifolium* are common, whilst *Drosera anglica*, *Myrica gale* and *Narthecium ossifragum* are absent. These drier habitats support a preponderance of bryophytes of drier, often woodland habitats, e.g. *Hypnum cupressiforme* and *Rhytidiadelphus loreus*.

The associations of the Sphagnion fusci are typical of the terminal, hummock phase of ombrotrophic mire development in which the ground water is relatively low in summer (0.2–0.3 m below the surface). Over its geographical range this alliance exhibits considerable variation and the dominant species of *Sphagnum* change markedly from west to east. *Sphagnum fuscum* predominates from Finland eastwards into Russia and south-eastwards through Germany into Poland, whilst *Sphagnum capillifolium* and *S. imbricatum* replace *S. fuscum* in more Atlantic areas. The character species of the alliance are *Rubus chamaemorus* (cloudberry), *Empetrum nigrum* (crowberry), *Sphagnum fuscum*, *S. capillifolium*, *Racomitrium lanuginosum*, *Dicranum undulatum*, *Cephalozia media*, *Calypogeia neesiana*, and the lichens *Cetraria islandica* and *Cladonia squamata*.

In western Ireland and north-west Scotland a variant of the Sphagnion fusci occurs in which *Racomitrium lanuginosum* (woolly hair-moss) forms hummocks up to 0.5 m high in the central raised plateaux of ombrotrophic mires, where it replaces *Sphagnum fuscum* and *S. capillifolium*. These tall hummocks are often colonised by *Eriophorum vaginatum*, *Calluna vulgaris*, *Deschampsia flexuosa* and several pleurocarpous mosses. Unlike the *Racomitrium lanuginosum* variant of upland blanket peat associations of the Ericion tetralicis, which is a retrogressive stage associated with peat erosion, the *Racomitrium* hummocks of the Sphagnion fusci are continuous into the peat surface to considerable depth and appear to be a true peat-forming vegetation type.

7.14 Peat-forming *Racomitrium lanuginosum* hummocks at Claish Moss, Argyll, Scotland.

Chapter 8

Salt marshes and sand dunes

These habitats are amongst the most environmentally severe of any in lowland Britain. They are subjected to extremes of wind, temperature and nutrient supply and very high concentrations of salt, either in sea spray or, in the case of salt marshes, tidal inundation. Diurnal and seasonal tidal movements also exert a dominant effect.

Coastal physiography is very diverse around the British Isles: in the north and west the old, hard rocks produce few opportunities for the development of coastal dunes and marshes, apart from in the estuaries of some large rivers, e.g. the Dee/Mersey and the Solway. Dunes and marshes are more frequently associated with the younger, softer rocks of eastern and southern Britain. Erosion of mineral debris from the land provides a reservoir of material for the development of sand dunes and salt marshes. Tidal action and coastal currents play a major role in the redistribution and subsequent deposition of this material. Wave action is responsible for moving mineral deposits back onshore and sorting the material into particles of different sizes. Three general size categories for these mineral deposits have been identified:

(a) silt (<0.1 mm diameter), which is transported in suspension and is very cohesive and resistant to scouring when compacted;
(b) fine sand (0.1–0.2 mm diameter), which is extremely mobile and may be transported in suspension; and
(c) coarse sand (>0.2 mm diameter), which travels as a unified bottom deposit.

The finer silts are the main substrate for salt marsh formation whilst the fine and coarse sands, aggregated as offshore sand banks, provide the wind blown sand necessary for sand dune development.

8.1 (a) View of salt marsh and sand dunes near Bettyhill, Sutherland. (b) General view over the Dee salt marshes on the Wirral, Cheshire. Salt marsh grass meadows (Puccinellion maritimae) are in the foreground with *Aster tripolium* (Asteretea tripolii) in the background.

8.1   Salt marshes

Maritime salt marshes have formed on many European coastlines, in the shelter of spits and islands and in protected, shallow bays. Around Britain there are large areas of salt marsh behind Blakeney Spit and Scolt Head Island, Norfolk; in the estuaries of the Rivers Humber, Thames, Ribble, Blackwater, Solway and Tay; and in Morecambe Bay, Southampton Water and the Wash. (Chapman, 1964).

The mud and silt banks of salt marshes are periodically flooded by sea water, which is the main agent of silt deposition. They are part of the littoral zone, which extends from the mean high water (MHW) mark to the mean low water (MLW) mark. The lowest regions of the marsh, the 'eu-littoral' zone, where the mud is very mobile, waterlogged and highly saline, are reached by the high water of neap tides (tides of smallest amplitude). Further inland, tidal inundation is less frequent and the highest levels of salt marshes, the 'supra-littoral' fringe, are only covered by the high water of spring tides (tides of greatest amplitude). The mud of the upper regions of salt marshes, although waterlogged, is more stable and less saline than that of the lower marsh. (Ranwell, 1972).

8.1.1   *Development of salt marshes*

Salt marshes can develop on rising, stable or submerging coastlines. Unless the sea is shallow, marshes of rising coastlines form in a narrow strip and if the rate of land elevation occurs relatively quickly, the salt marsh is soon replaced by freshwater marsh or damp grassland. On stable coastlines salt marsh development is governed by the slope of the sea shore and the rate at which deposition of mud, and its subsequent erosion, take place. The net deposition of silt on a sinking coastline obviously must exceed the rate at which land levels are falling, otherwise salt marsh will not develop.

The formation of salt marsh is dependent on accumulation of mineral material of small particle size: silt or mud transported downstream by rivers and deposited on the banks and shallows of river estuaries, or sediments brought down to the coast by rivers and, subsequently, carried along the coastline by tidal action and deposited elsewhere. In addition, there may be some deposition of wind-blown sand carried by onshore winds. The rate at which the mineral material of salt marshes accumulates depends on a number of factors:

(a) the extent of the tidal reach – the greatest rate of deposition occurs on the lowest levels of the marsh, which are covered by tides for the longest daily period;

(b) vegetation cover – this removes mineral material from suspension in sea water as the rate of water flow decreases;

(c) tidal directional movements – redistribution of sediments is brought about during changes in offshore currents;

(d) the high sodium content of sea water, which induces flocculation of soil colloids which later sediment; and

(e) the degree of dilution of sea water by fresh water, which affects the sedimentation rate of silt and mud.

Deposition and accrual of sediments can increase the height of the mineral ground above the marine influence and induce successional changes to take place. Erosion can also occur, with the opposite effect.

8.1.2   *Special features of the salt marsh environment*

Apart from the overriding influence of twice-daily tidal movements over and around salt marshes, several other physical and chemical features are of great importance, including water table, aeration and salinity.

8.1.2.1   *Water table*

The water table of salt marshes fluctuates in response to tidal movements. It is also influenced by soil structure: where there is a high proportion of sediments of small particle size (i.e. clay and silt) the marsh will have poor drainage properties; where the proportion of sand is higher, drainage is improved. The extent to which different parts of the salt marsh system are waterlogged during the day, and the manner in which they are able to drain away the salt water as the tide level drops, have a great influence on the plants which grow there.

8.1.2.2   *Soil aeration*

The oxygen content of salt marsh soils increases from the lower reaches of maximum inundation to the more stable, elevated regions which are only occasionally exposed to tidal movements. However, even when flooded by the highest tides, the ground water table in salt marsh soils never rises completely to the surface because some air always remains trapped. This aerated layer, which corresponds to the rhizosphere for most species, is very important for the survival of plants which grow in the lower regions of salt marshes because root respiration is never completely impeded. The high oxygen levels of the upper layers of salt marsh soils also facilitate decomposition of much of the annual organic production of plants and animals, and most of the remainder is exported by the tide, so that there is negligible accumulation of organic matter.

8.1.2.3 *Salinity*
The overriding factor influencing the vegetation of saline habitats is the concentration of sodium and chloride ions in solution. The amount of sodium chloride in a substrate depends on the duration of tidal flooding, height of the ground water table, leaching effect of rainfall, physical nature of the soil, slope, and the ameliorating effect of fresh waters. Interaction between these factors and the variations in salinity that they produce leads to the marked zonation of vegetation that is characteristic of salt marshes and other coastal habitats. Plants of saline environments (halophytes) have anatomical and physiological adaptations which ensure their survival.

8.1.2.4 *Salt tolerance*
Halophytes obtain their water and mineral nutrients from a soil environment in which there is a high concentration of soluble elements. On the basis of physiological differences in their responses to salinity, halophytes are classified into either salt-regulating or salt-accumulating species. The former restrict the internal accumulation of salts by reducing their uptake from the soil (e.g. *Atriplex* spp.), whereas the latter accumulate salts to high internal concentrations and extrude excess through specialised salt glands on to their external surfaces (e.g. *Spartina townsendii*). (Flowers *et al*, 1977).

8.2   Vegetation of saline soils

A distinct zonation of plant communities occurs on coastal salt marshes. This zonation, which corresponds to the salt marsh succession or 'halosere', develops in response to different daily periods of inundation by the tide. It extends from pioneer vegetation of the permanently waterlogged and unstable sand and mud of the eu-littoral zone, to the drier, elevated and stable salt marsh meadows or 'flats' of the littoral fringe, where submergence only occurs during the highest tides of the year. A succession of vegetation types occurs in response to rising land levels produced by tidal silting, and, on

those areas most remote from tidal influence, organic matter accumulates.

8.2.1   *Marine grass communities of estuaries and mud flats (Zosteretea)*

These submerged associations are widely distributed around British coasts in shallow sea water (to a depth of 3–4 m) on shingle and sand in the sub- and eu-littoral zones. The character species are members of the genus *Zostera*, individual species of which have different habitat preferences. *Zostera marina* (eel-grass), which is very susceptible to tidal movements, occurs on sandy mud in the sub-littoral zone from the middle tidal reach

8.2 Pioneer *Spartina* × *townsendii* vegetation (Spartinetea) being replaced by supra-littoral vegetation of the Asteretea tripolii as salt marsh mud accretion raises the level above the normal daily tidal reach.

to a depth of 4 m. *Zostera angustifolia* (narrow-leaved eel-grass) grows with various marine algae, mainly from the high tide line to a depth of about 1.5 m, where occasionally it forms a vegetation mosaic with *Zostera noltii* (dwarf eel-grass).

8.2.2   *Pioneer communities of waterlogged saline and muddy substrates (Thero-Salicornietea)*

This therophytic vegetation forms fairly dense swards in the muddy delta region of salt marshes, often in close proximity to *Spartina* × *townsendii* (Townsend's cord-grass). The character species of the class, the order Thero-Salicornietalia and the alliance Thero-Salicornion, is *Salicornia europaea* (glasswort), a small, succulent species which colonises mud, clay and silt to a depth of 30 cm below the MHW line. It is distributed throughout northern and western Europe and is common around the entire British coastline.

8.2.3   *Cord-grass salt marshes (Spartinetea)*

The genus *Spartina* forms single species stands of pioneer vegetation on low-lying, waterlogged mud, clay and sand in the eu-littoral zone, often replacing associations of the Zosteretea. *Spartina maritima* (small cord-grass) colonises tidal mud flats, on which brown and green algae may be abundant. It often occurs with *Salicornia europaea* in the middle high water zone. The invasive hybrid *Spartina* × *townsendii* achieves optimal growth on wet mud in the eu-littoral zone, where it is more resistant to tidal erosion than other pioneer plant associations. It is an allopolyploid hybrid of *S. maritima* and *S. alterniflora* (smooth cord-grass) which was first discovered towards the end of the nineteenth century in Southampton Water and has since spread along the English and north-western European coastline. It occurs in the delta region of estuaries from 1 m below to about 0.1 m above MHW and is often planted to stabilise mud flats and to prevent erosion.

8.2.4   *Grazed associations of the supra-littoral zone (Asteretea tripolii)*

Associations of this class are the grass and herb-rich vegetation of the upper region of salt marshes. The single order Glauco-Puccinellietalia contains four alliances: Puccinellion maritimae, Armerion maritimae, Puccinellio-Spergularion salinae and Halo-Scirpion, the species of which collectively colonise the elevated zone between the maritime edge and the storm tide upper limit, and also the margins of water channels which dissect salt marshes. The character species of the class are *Aster tripolium* (sea aster), *Plantago maritima* (sea plantain) and *Triglochin maritima* (sea arrow-grass).

8.2.4.1   *Salt-marsh grass meadows (Puccinellion maritimae)*

These associations form short, grey-green swards covering muddy sand; they are frequently inundated by storm and spring high tides. Character species include *Puccinellia maritima* (common salt-marsh grass), *Cochlearia anglica* (English scurvy-grass), *Halimione portulacoides* (sea purslane), *Limonium vulgare* (common sea lavender), *Plantago maritima* and the red algae *Bostrychia scorpioides* and *Catanella opuntia*.

8.2.4.2   *Herb-rich coastal meadows (Armerion maritimae)*

Associations of this alliance form low-growing lawns on silty or muddy ground from 0.2 m below MHW to the maximum reach of storm tides; they may succeed vegetation of the Puccinellion maritimae and occur on most coastal marshes around the British coastline. Characteristic species are *Armeria maritima*, *Glaux maritima* (sea milkwort), *Juncus maritimus* (sea rush), *Juncus gerardii* (salt-marsh rush) and the grasses *Festuca rubra*, *Alopecurus bulbosus* (bulbous foxtail) and *Agrostis stolonifera*. *Juncus maritimus* often occurs in great abundance and this species, together with *J. gerardii*, forms dense swards on heavily grazed coastal meadows, especially on drained and reclaimed salt marshes.

8.2.4.3   *Tidal channels and creeks (Puccinellio-Spergularion salinae)*

These are ephemeral associations of dry channels in the upper reaches of salt marshes which are periodically inundated by storm tides and, rarely, inland on sandy saline soils near brine springs. *Puccinellia distans* (reflexed salt-marsh grass) and *Spergularia marina* (lesser sea spurrey) are characteristic, but *Atriplex prostrata* (spear-leaved orache) and *Parapholis strigosa* (hard-grass) may also occur.

8.2.4.4   *Sea club-rush community (Halo-Scirpion)*

This alliance is characterised by *Scirpus maritimus* (sea club-rush), *Atriplex prostrata*, *Triglochin maritima* and *Aster tripolium*. It occurs on sandy substrates flushed

8.3 Different stages in the development of a sand dune system: (a) embryo dune formation at Aberffraw, Anglesey; (b) strand and mobile dunes at Newborough, Anglesey; (c) semi-permanent dunes and dune slacks at Newborough.

with fresh water from the landward side of salt marshes and as a result it also contains some species typical of the Phragmitetea, e.g. *Phragmites australis* (reed-grass) (section 6.5.4).

8.2.5   *Brackish pools and ditches (Ruppietea)*

The vegetation of this class occurs locally around the coast of the British Isles in the brackish waters of coastal ponds and ditches which are only inundated by the highest tides, but where there is a strong influence of sea-spray. The character species of the class, the order Ruppietalia, and the single alliance Ruppion maritimae is *Ruppia maritima* (beaked tasselweed), which frequently grows with *Zannichellia palustris* (horned pond-weed) and, occasionally, *Ruppia cirrhosa* (spiral tasselweed) and the rare *Eleocharis parvula* (dwarf spike-rush).

8.3   Sand dunes

Sand dunes develop in coastal areas (but occasionally inland), as a result of the accumulation of sand blown onshore by strong prevailing winds. This sand originates in deposits which have formed offshore by a combination of erosion of terrestrial sandstone rocks and wave action, which forms sandbanks in coastal regions. At low tide, these shoals are exposed and, as their surface dries, sand grains are blown onshore where they are redeposited (Fig. 8.4). The sand dune habitat is inherently unstable and involves processes of accretion and erosion. The movement of sand across dunes occurs by saltation: sand particles carried onshore fall to the dune surface under gravity and bounce back into the wind, thereby disturbing other grains on the unstable sand surface, setting

them in motion. Sand grains are trapped and removed from the air when they encounter obstacles in their path, forming embryo dunes, the size of which is limited by the height of the obstruction. Any obstacle which is elevated above the generally flat surface beyond the high tide zone will serve as a site for dune initiation, e.g. flotsam and seaweed washed up by the tides, or the sparse vegetation of the foreshore. (Ranwell, 1972)

Embryo dunes are characteristic of the strand (i.e. that part of the shore between the limit of normal high tides and that reached by the highest storm tides) of sandy beaches, where they are continually being formed and removed as a result of wind action or occasional inundation by the tide. On the seaward side of the embryo dune zone, constant movement of the sand by waves and currents twice daily prevents dune formation and plant colonisation – even the larger algae are absent. Embryo dunes develop an aerofoil shape because the lighter grains which are carried up the windward side of the dunes slide down and settle on the less steep, leeward slope. These dunes provide a habitat for the establishment of perennial grasses, including *Elymus farctus* (sea couch) and *Leymus arenarius* (lyme-grass), which reduce the wind speed just above the dune surface, cause further blown sand to accumulate and are, therefore, partly responsible for the limited increase in height of these low dunes.

The growth of pioneer grasses on embryo dunes is restricted by their relatively shallow root systems, which limit water supply to the plants; by frequent, and often sudden, burial by new deposits of wind-blown sand; and by occasional inundation by storm tide water. Embryo dunes are periodically blown away and this sand contributes to the formation of the larger, semi-permanent mobile dunes or yellow dunes further inland.

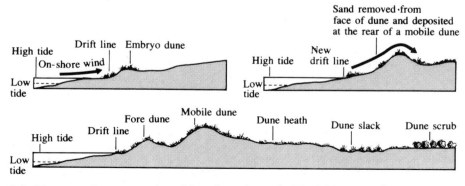

8.4   Diagrammatic representation of dune formation and of the habitats on a dune system.

The dominant species of mobile dunes is *Ammophila arenaria* (marram grass), a tough, invasive perennial with a rhizome capable of almost unlimited horizontal and vertical growth which pushes new shoots up to the sand surface following burial, enabling it to keep pace with rapid dune enlargement. The rhizomes of these plants form a dense network which permeates the whole dune hill. As dunes become more stable, other species colonise the sand (including mosses and lichens) and dune-building species, such as *Ammophila arenaria*, die out. Humus accumulation in the surface layers of these stabilised, fixed dunes gives them a darker colouration, hence the alternative term 'grey dunes'; further stages in dune succession comprise scrub and heathland development.

The maximum height of the mobile dunes is limited by the effectiveness of the upward growth of *Ammophila arenaria* and by the degree of exposure to wind. The highest dunes are those furthest inland when the prevailing wind is onshore and nearest the coast when it is offshore. In the latter, which mainly occur along the east coast of Britain, sand dunes only occupy a narrow coastal strip or form offshore islands or spits. (Ratcliffe, 1977).

The foredunes (embryo and mobile dunes) are susceptible to change, especially during winter gales, when large, additional depositions of sand can lead to reduced vigour and even death of the colonising plants at a time when growth is at a minimum, or has ceased altogether. Strong winds may remove large sections of sand dunes by 'blowing out' the sand which is redeposited still further inland, so that over a period of time the dune system effectively moves further onshore (Chapman, 1964). Continual development of new series of embryo and mobile dunes in front of those already established provides some protection to the older dunes which are blown away less frequently than those nearer to the sea, although the formation of new dune ridges to the seaward side is restricted by the undercutting action of high tides. As a result, more than one series of dunes may be produced parallel to the shore, with alternating ridges and hollows in between. The rate at which the sand moves is dependent on wind velocity and dune height: embryo dunes and small mobile dunes migrate the fastest. The mass movement of sand inland can create problems: before planting by the Forestry Commission in the 1920s, the Culbin dunes (on the Moray coast of north-east Scotland) were the most extensive in Britain. However, progressive inundation of farmland and habi-

tations by sand (including the village of Culbin) resulted eventually in the establishment of large plantations of conifers to stabilise this stretch of coastline. Similar loss of settlements and agricultural land by the landward movement of dunes has also occurred along parts of the Norfolk coast.

The low-lying depressions between dune ridges are called slacks, and these may be either wet or dry depending on the water content of the substrate and the maximum depth to which the water table falls during the year. In wet slacks the water table never falls below 1 m throughout the year and during winter and spring these slacks are frequently flooded. This inundation, if it persists for long enough, can lead to colonisation by aquatic vegetation (Ch. 6). The water table in dry slacks is always 1–2 m below the surface and the vegetation is xeromorphic.

8.3.1 *Sand dune soils*

Sand dune soils are regosols, which have no horizon development. The surface sand has a low humus content and water-holding capacity and nutrients are rapidly leached from the upper layers. Dune sands vary in chemical composition from siliceous sands derived from weathered sandstone rocks to very calcareous sands consisting mainly of calcium carbonate from the comminution of mollusc shells. The water table (usually more than 2 m below the dune surface) is normally too low to supply water to plant roots by upward capillary movements. The main sources of water to plants, are therefore, rainfall and, to a lesser extent, dew, which may be retained by stray organic material near the surface or absorbed by the roots of colonising plants. Dune slacks, since they are nearer to the water table than dune hills, usually have a geogenous water supply.

The properties of dune soils change with time under the influences of increased stability and downward movement of rainwater through the sand: the progressive accumulation of plant and animal remains increases the surface organic content, whilst leaching by rainwater reduces the calcium carbonate content (and hence the alkalinity) of the surface layers, i.e. the pH decreases.

8.3.2 *Adaptations to the dune environment*

The main problems for the plants which survive in this ecosystem are: instability of the substrate as a result of constant wind action, which may either remove or bury

a

b

8.5 (a) View of *Ammophila arenaria* community (Ammophiletea). (b) *Ammophila arenaria*.

the vegetation; lack of soil moisture; and excessively high concentrations of certain ions and a deficiency of others, including several required in plant mineral nutrition.

8.3.2.1 *Plant growth strategies*

Ammophila arenaria and *Elymus farctus* have negatively geotropic rhizomes which maintain their shoots at the surface by rapid upward growth. Rhizome development and adventitious root production ensure stability and also exploitation of a large soil volume for water acquisition. The rhizome apex of *Ammophila* is hard and sharply pointed, which enables it to move easily and without contusion as it rapidly pushes through the dune sand (Fig. 8.5). Other species such as *Eryngium maritimum* (sea holly) and *Euphorbia paralias* (sea spurge) are also capable of rapid upward growth following burial by sand. (Salisbury 1952).

Although bryophytes are not capable of the same rapid rate of growth, some moss species can survive inundation by sand to a depth of several inches. *Bryum algovicum* var. *rutheanum* produces new shoots after burial and the dense cushions which this moss forms assist in the sand stabilisation process.

Many of the plants that colonise the drift-line and strand are annuals (therophytes) which can quickly re-colonise from seed following the frequent disturbance caused by the actions of both wind and tide. The flora of stabilised mobile dunes also includes a large number of winter annuals, which germinate in the autumn months, and flower in the spring or summer, thus restricting their growth to the damper periods of the year and avoiding the arid summer months.

8.3.2.2 *Water balance*

As a result of their free-draining nature and low colloidal content, young dune sands have a very low water-holding capacity. In older dune soils, the water relations are improved as a result of their higher organic content.

Salisbury (1952) recorded minimum soil water contents at Newborough Warren, Anglesey, of less than 1% and he stressed that water is unavailable to plants below 0.5%, the point at which wilting ensues. However, many plants of sand dunes have xeromorphic adaptations to reduce water loss and, thereby, increase their chances of survival in this unfavourable environment, e.g. *Ammophila arenaria* has narrow, cutinised leaves which inroll to protect the stomatal surfaces (Fig. 1.2). Several plants of dunes, particularly young dunes, are

succulents, e.g. *Salsola kali* (prickly saltwort), and have fleshy leaves with well-developed water-storage tissues.

The rooting systems of dune species are of two types: shallow roots which can take up moisture from the surface layers of sand from precipitation, and deep roots which tap layers of moist sand within the dune hill.

8.3.2.3 *Nutrient balance*

With the exception of calcium in calcareous dunes, coastal sands are extremely nutrient-poor and contain only small amounts of most of the mineral salts essential for plant nutrition, particularly nitrates and phosphates. However, the presence of nitrogen-fixing bacteria and blue-green algae (Cyanobacteria) in dune soils may supplement the nitrogen economy of the dune system, particularly during the early stages of dune development. The rhizosphere of several dune grasses (e.g. *Elymus* spp. and *Ammophila arenaria*) contains nitrifying bacteria which are only present in low numbers elsewhere in the sand, and *Hippophae rhamnoides* (sea buckthorn), a shrub of the nutrient-deficient, highly leached soils of stabilised dunes, has root nodules which contain nitrogen-fixing bacteria (actinomycetes). Some species have a high nitrogen requirement (nitrophiles) and are often found growing on decomposing organic material of marine or terrestrial origin. Several plants whose seeds are distributed with the drift are of this type (e.g. *Atriplex* spp.), and their germination at the drift-line ensures a good supply of nitrates. The increased organic content of the sand at the drift-line also improves its water-holding capacity. There is also evidence to suggest that dune slack sedges (especially *Carex flacca*) can obtain phosphorus by the cleavage of inorganic phosphorus directly from organic substrates by the exudation of the enzyme phosphatase into the rhizosphere in the vicinity of specialised, swollen, lateral roots (Davies *et al.*, 1972). This may promote plant growth in dune substrates which are phosphorus-deficient.

In spite of their proximity to the sea, coastal sands are almost completely devoid of sodium chloride, which is leached out at an early stage in the dune successional sequence. High concentrations of sodium are, therefore, only a problem for the few pioneer species of the embryo dunes nearest to the sea. Elsewhere, high concentrations of calcium and/or a deficiency of most other ions are the main problems for plant growth. The cation exchange capacity of dune soils is low, owing to the minimal clay and organic matter contents, and this limits the supply of cations and water to plant roots; pH can also exert a strong influence on nutrient availability in dune soils and, in addition, it has an effect on the solubility of certain ions (e.g. aluminium, a phytotoxin, becomes increasingly soluble at low pH, whereas high pH can lead to a decrease in the availability of iron, potassium and phosphorus).

8.3.3 *Vegetation of sand dunes*

Distinct zonational changes in plant communities can be observed from the coastal to the landward side of a dune system. The embryo dunes, with their scanty plant cover, are succeeded by mobile and eventually stabilised grey dunes, which have an almost complete vegetation cover. Within the different stages of this system, small-scale variations in topography (which influences aspect and exposure) result in a range of local microclimatic conditions. Small-scale vegetational differences may also result from the presence of a localised nutrient source, e.g. rotting seaweed on the strand or deposits of seagull guano.

In addition to the pioneer and colonising associations of strand, embryo and mobile dunes of the Cakiletea maritimae and the Ammophiletea, plant communities of other classes form an important component of sand dune ecosystems. At the landward side of dunes the stabilised sand supports low dune scrub of the class Rhamno-Prunetea (section 8.3.3.4), or grassland vegetation of the class Molinio-Arrhenatheretea (section 5.3.2) where the calcium carbonate content of the soil is still high, or heathland vegetation of the class Nardo-Calluneta (section 5.4) where the soil is more heavily leached and has a surface layer of acid, mor humus. Occasionally, where topographical features lead to water impoundment, or an impervious substratum prevents drainage, wetland vegetation of the classes Phragmitetea and Potametea (sections 6.5.2 and 6.5.4) can develop, which, if succession proceeds for long enough, may develop into ombrotrophic mire vegetation of the Scheuchzerietea and Oxycocco-Sphagnetea (sections 7.3.1 and 7.3.2).

8.3.3.1 *Strand and maritime edge vegetation (Cakiletea maritimae)*

The associations of this class are fragmentary and are strongly influenced by salinity and substrate disturbance. The character species is *Atriplex prostrata* (spear-leaved orache), which, in common with many other species of

this class, is a therophyte. There are two orders: Thero-Suaedetalia and Cakiletalia maritimae.

The Thero-Suaedetalia contains pioneer associations of the zone immediately above the normal daily tidal limit on sand with a high proportion of decaying organic material (i.e. a high nitrate content). The principal species is *Suaeda maritima* (annual sea-blite), but on the seaward side there may be a gradation into salt marsh vegetation of the Spartinetea (section 8.2.3).

The Cakiletalia maritimae contains several pioneer associations of the strand and embryo dune regions of the foreshore, characterised by *Cakile maritima* (sea rocket), *Salsola kali*, *Atriplex littoralis* (grass-leaved orache) and *A. prostrata* all of which are annuals. The *Atriplex* spp. colonise seaweed debris and other organic material which has been deposited above the tidal reach, whilst occasional patches of vegetation dominated by *Crambe maritima* (sea kale) may occur on sand or shingle. Above the normal tidal limit, in regions frequently overblown by dry sand, low-growing pioneer vegetation dominated by *Honkenya peploides* (sea sandwort) and *Salsola kali*, and often containing *Cakile maritima* and *Glaucium flavum* (yellow horned poppy), occurs. These plants may be responsible for the initiation of low, embryo dune formation.

8.3.3.2 *Vegetation of sand dune hills (Ammophiletea)*

The associations of this class are dominated by rhizomatous geophytes and ephemeral herbs. The character species are *Ammophila arenaria* (marram), *Honkenya peploides*, *Calystegia soldanella* (sea bindweed), *Euphorbia paralias* and *Eryngium maritimum*. The single order Elymetalia arenarii contains two alliances: Agropyro-Honkenyion peploidis and Ammophilion borealis.

The Agropyro-Honkenyion peploidis embraces the pioneer associations of predominately shallow-rooted species of the foredunes and embryo dunes. This alliance is characterised by the rhizomatous grass *Elymus farctus* (formerly *Agropyron junceiforme*). Vegetational variation within this alliance arises in response to the degree of exposure to the highest storm tides: on embryo dunes which are only flooded by the highest tides *Euphorbia paralias* and the procumbent *Calystegia soldanella* are abundant, whilst just beyond the tidal limit the grass *Leymus arenarius* often predominates.

The mobile dunes which are colonised by *Ammophila arenaria* and other deep-rooted species are in the alliance Ammophilion borealis. These communities play

8.6 Sand dune species: *Eryngium maritimum*, an associate of *Ammophila arenaria*.

an important role in stabilising the coastal environment, and *Ammophila arenaria*, in particular, has been extensively planted in attempts to bind and consolidate drifting sand. The temporary stability effected by this grass enables other species to colonise the dunes. These include *Eryngium maritimum*, *Leymus arenarius* and *Festuca rubra* subsp. *arenaria* (red fescue). Eventually the area of bare sand decreases as the dune successions moves towards the fixed dune stage and other plants, in particular mosses (e.g. *Bryum* spp. and *Tortula muralis*) and lichens (principally species of *Cladonia* and *Peltigera*), start to colonise the sand surface. A variety of small, annual species are also characteristic of the fixed dune flora, e.g. *Sagina procumbens* (procumbent pearlwort), *Stellaria media* (common chickweed), *Trifolium* spp. (clovers) and *Potentilla anserina* (silverweed). In common with other plants of this part of the dune system, several of these are 'adventives', i.e. species introduced directly or indirectly by man.

8.3.3.3 *Vegetation of dune slacks (Nardo-Callunetea, Salicetea purpurea and Parvocaricetea)*

The sheltered environment of low-lying, dry dune slacks provides a suitable habitat for the development of a low-growing vegetation dominated by *Salix repens* (creeping willow) of the order Calluneto-Ulicetalia, class Nardo-Callunetea. This occurs on dry substrates which have a well-developed O horizon of mor humus. Other charac-

teristic species are *Pyrola minor* (common wintergreen) and less commonly *Pyrola rotundifolia* (round-leaved wintergreen), *Monotropa hypopitys* (yellow bird's-nest) and *Gymnadenia conopsea* (fragrant orchid).

In damp dune slacks where both the calcium and organic content of the surface sand is high, and the ground water level reaches the surface in winter, a low-growing, shrubby vegetation dominated by *Salix purpurea* (purple willow) and *S. repens* frequently occurs, in which *Betula pubescens*, *Salix aurita × cinerea* and the small herbs *Hydrocotyle vulgaris* (marsh pennywort), *Prunella vulgaris* (self-heal), *Eleocharis palustris* (common spike-rush), *Pulicaria dysenterica* (common fleabane) and *Moehringia trinervia* (three-nerved sandwort) are common. This vegetation belongs to the class Salicetea purpureae.

Coastal vegetation of the class Parvocaricetea occurs wherever the water table is high for most of the year in the depressions between the dune hills and where on occasion there are areas of standing water. Depending on the nature of the substrate sand, the pH may be alkaline or acid and this determines the species comsition, which may resemble the Tofieldietalia if calcareous or the Caricetalia nigrae if acid (sections 6.5.6.1 and 6.5.6.2). Species common to both orders include the sedges *Carex arenaria* (sand sedge) and *C. flacca* (glaucous sedge), plus *Hydrocotyle vulgaris*, *Mentha aquatica* (water mint), *Ranunculus flammula*, *Juncus articulatus, Equisetum palustre* and *Salix repens*. In calcareous dune slacks the species diversity is greater and additional species may include *Carex serotina* (small-fruited yellow sedge), *Parnassia palustris* (grass of Parnassus), *Equisetum variegatum* (variegated horsetail), *Schoenus nigricans* (black bog-rush), *Linum catharticum* (fairy flax), *Pinguicula vulgaris* (common butterwort) and the bryophytes *Campylium stellatum, Bryum pseudotriquetrum, Scorpidium scorpioides* and *Pellia fabbroniana*.

8.3.3.4 *Dry scrub vegetation of stabilised dunes (Rhamno-Prunetea)*

The associations of this class result largely from grazing by sheep and/or rabbits, stabilisation of dunes by planting, or damage by human visitors. Many of the communities are temporary and result from degradation of the dunes.

Characteristic of the alliance Berberidion (order Prunetalia spinosae), is the shrub *Hippophae rhamnoides* (sea buckthorn), which forms a dominant vegetation on stabilised dunes along the east coast of Britain. Where the sand dunes have been little disturbed the vegetation may be rich in species and include *Rosa pimpinellifolia* (burnet rose), *R. canina* (dog rose), *Rhamnus catharticus* (buckthorn), *Crataegus monogyna* (hawthorn), *Inula conyza* (ploughman's spikenard) and *Lithospermum officinale* (common gromwell). Elsewhere the species diversity is much less and *Sambucus nigra* (elder) may be co-dominant. Frequently present in the latter habitat may be *Stellaria media* (common chickweed), *Anthriscus caucalis* (bur chervil) and *Bromus sterilis* (barren brome).

Occasionally, communities of the alliance Salicion arenariae colonise dry, calcium-deficient dune slacks on the coasts of East Anglia and Lincolnshire. This vegetation occurs in close proximity to the nitrogen-fixing *Hippophae rhamnoides* scrub, and *Salix repens, Fragaria vesca* (wild strawberry), *Avenula pratensis* (meadow oat-grass), the ferns *Polypodium vulgare* (common polypody) and *Botrychium lunaria* (moonwort) and the feather-moss *Hylocomium splendens* are frequent.

Chapter 9

Influence of altitude and latitude on vegetation

Significant environmental changes occur on moving from low to high altitudes and from middle latitudes to polar regions. In general, habitat conditions become more severe and, as a result, the vegetation contains a large number of plants with morphological and physiological adaptations which enable them to grow and reproduce under the harsh environmental conditions and to recover quickly from periods of environmental stress. However, increases in altitude and latitude, although greatly influencing the growth, development and survival of plants, cannot be regarded as primary environmental factors. Rather, they affect vegetation indirectly through changes induced in the climatic, edaphic and biotic influences which control plant performance and distribution.

A fundamental difference between high altitude or polar regions and middle latitude or low altitude regions is the length and nature of the growing season. In temperate zones this usually lasts for six months or more, during which time plants grow vigorously to complete their growth cycle, often with considerable increases in biomass; the unfavourable winter period is commonly avoided through dormancy. In contrast, the growing season in montane and polar regions is short and the environment inimical to growth. Although some plants occur in both montane and polar regions the two environments differ significantly with respect to: light quality and quantity; duration of frozen soil and snow cover; wind velocity; and the partial pressures of oxygen and carbon dioxide. The British flora contains plants which occur in both arctic and alpine environments (section 9.3).

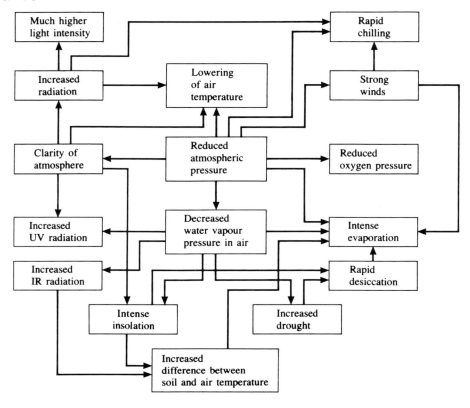

9.1   The relationships between the different climatic factors in mountains (after Dajoz, 1977).

9.1   Environmental factors of high altitudes and latitudes

The rigours of mountain and northern environments impose severe restrictions on vegetation and, in particular, climatic factors limit the composition of the vegetation and the species distribution. However, the availability of ground water or snow-melt water, and the pH and nutrient content of the soil are also important (Fig. 9.1).

9.1.1   *Light*

Daylength varies with change in latitude. At the equator there are 12 hours of daylight and 12 hours of darkness every day of the year. With increase in latitude northwards or southwards daylight alters to more than 12 hours in summer and less than 12 hours in winter. Beyond 66° latitude the sun shines for 24 hours a day in mid-summer, whereas in mid-winter only a faint, scarcely detectable light is in evidence at noon. These large seasonal variations in daylight between different latitudes are unaffected by increases in altitude.

Solar radiation is less attenuated by the rarified atmosphere of high elevations than in lowland situations and at high altitudes the blue and ultraviolet parts of the spectrum comprise a large proportion of the incident radiation. This effect is most marked on high mountain ranges, whilst at lower altitudes, for example on the hill-lands of Britain, it is often mitigated by the presence of cloud.

9.1.2   *Temperature*

In general, average annual temperatures decline on moving from the equator to the poles and as land elevation increases. However, geographic variations in temperature are often moderated by the proximity of large inland lakes or the sea, an aspect that has a profound influence on the climate of the British Isles. Air temperatures decrease by an average of 5.5°C for every

1000 m increase in height (a value that varies with latitude and season of the year). This altitudinal temperature gradient is greater on the lower slopes of mountains than at higher elevations, and on north-facing slopes than on those facing south (vice versa in the southern hemisphere).

Diurnal temperature fluctuations are less in northern than mountain regions, where maximum daily temperatures in excess of 25°C can be recorded in summer, especially close to the ground. The less attenuated infra-red and ultraviolet radiation experienced by high mountain regions can produce a ground surface temperature 10–20°C higher than that of the air above. However, re-radiation of much of this heat at night can cause frosts even during the summer months.

9.1.3 Water

Rainfall varies considerably with latitude. Potential rainfall is highest in equatorial regions and lowest in polar ones, where the air is too cold to hold much moisture. Mountain alpine tundras receive about 100–200 cm of rain annually compared to a precipitation of only 10–15 cm in polar regions. Up to about 1500 m above sea level (a height which encompasses all British mountains), rainfall increases with altitude as a result of the upward deflection of moisture-laden onshore winds by mountain ranges. On rising, this damp air is cooled, water vapour condenses and, if there is sufficient quantity, falls as precipitation. At higher altitudes this relationship ceases, since most of the rain has already fallen on the lower slopes, and the colder air temperature induces the remainder to fall as snow.

9.1.4 Snow

At high altitudes and latitudes the extent of snow cover varies considerably, ranging from a minimal quantity that melts early in the season, to deep drifts that persist throughout the summer, particularly on north-facing slopes. Snow cover has an important selective effect upon plants, seriously damaging some whilst benefiting others. If cover is prolonged those plants which are able to complete their life-cycle during a short, snow-free period are favoured.

9.1.5 Carbon dioxide supply

Along with the drop in atmospheric pressure and partial pressure of water vapour, partial pressures of oxygen and carbon dioxide also decrease with altitude. The pressure of carbon dioxide, for example, at 1000 m is 0.89 bar; at 4000 m it is 0.60 bar. A low concentration of carbon dioxide may be stressful to plants.

9.1.6 Wind

Strong winds limit plant growth at high altitudes and latitudes, an effect that is usually greater on slopes facing the prevailing wind. However, wind patterns vary greatly with topography, and obstructions on the ground, such as stones, rocks and vegetation may create sheltered microhabitats in which susceptible species can survive.

9.1.7 Aspect

In mountain areas striking differences in microclimate occur as a result of aspect (direction of slope). At highest elevations this effect is so extreme that the minimum soil temperature on south-facing slopes may be higher than the maximum on north-facing ones. A slope of as little as 5° towards the pole reduces soil temperature by as much as an increase of 10° of latitude in the same direction. As a consequence of this slope-insolation relationship many plants of warmer, drier lowland habitats grow in mountain regions on slopes and ridges which receive maximum amounts of solar radiation. In contrast, plants adapted to the cool, moist environments of high altitudes extend downwards to lower elevations in shaded, damp ravines on cold, north-facing slopes.

9.1.8 Soils

In highland and northern Britain the landscape and soils are largely determined by the widespread distribution of hard rocks – granite, quartzite, sandstone, non-calcareous schists and gneisses. The soils to which these give rise are often juvenile, unstable and, even if developed to maturity, are leached, nutrient-deficient podzols. The only source of nutrient enrichment to these acid soils is where the surface is flushed with water (from springs or snow-melt), or where the influence of gravity or frost reduces the stability of the soil and causes churning of the surface layers. Where waterlogging occurs a superficial layer of peat will often develop. Soils of higher nutrient content are associated with limestone and dolomitic rocks, calcareous mica schists, calcareous

9.2 Solifluction ridges on Ben Lawers, Perthshire, where the steep slopes encourage surface soil movement downslope.

9.3 The cushion growth form of *Silene acaulis* enables this plant to survive in the harsh environment of montane and northern regions.

basalt and serpentine. Soils derived from these richer rocks are less widely distributed in upland northern Britain than are acidic ones, although many regions of acid rock contain outcrops of calcareous strata which give rise to localised nutrient-enriched substrates. Owing to the steepness of mountain slopes and the cold and wet climate, solifluction is widespread.

9.2 Plant modifications to mountain and northern environments

9.2.1 *Growth form and morphology*

The most obvious modifications to plants imposed by the severe conditions resulting from increases in altitude and latitude are a reduction in height and a tendency to adopt a perennating life-form. Trees and shrubs become increasingly stunted by high winds and are replaced by lower-growing dwarf shrubs and herbs in extreme situations. This 'tree-line' is either a latitudinal limit, for example, the boundary between tundra and taiga (sections 4.1 and 4.2) or an altitudinal one on mountainsides where sub-montane forest gives way to montane heath. The altitudinal limit of the tree-line varies with latitude, decreasing by approximately 110 m for every degree of latitude northwards. In Britain the tree-line probably lies at about 650 m, although the natural limit to tree growth is no longer in evidence as a result of forest removal, burning, grazing and re-afforestation. Maritime proximity can also exert a modifying influence and there is an increase in the altitude of the tree-line with distance inland. The tree-line decreases down to sea level in the north of Scotland.

Above the tree-line the majority of herbaceous species have rosettes, cushions or short, leafy stems. Shrubs, which are usually less than 15 cm tall, produce little additional woody tissue each year and seldom reproduce sexually – recolonisation, species maintenance and distribution being achieved more successfully by vegetative means. Low-growing monocotyledonous plants (grasses, sedges and rushes) are common and these have apical meristems near to the ground where they are protected from the harsh climate by a covering of soil and vegetation. The dwarfing of plants ensures that growth in summer takes place in the warmest zone close to the soil surface. At intermediate altitudes and latitudes, where duration of snow cover is short and the soil does not freeze for long periods, many plants overwinter by underground perennating organs – swollen root stocks, rhizomes, corms, bulbs or tubers – in which carbohydrate food reserves are stored through the winter and mobilised early the following year when temperatures start to increase again. In colder climates, where the subsoil remains frozen for much of the year, shrubs may retain considerable carbohydrate reserves in their stems and leaves.

Xeromorphy and the evergreen habit are common adaptations in flowering plants of arctic and alpine environments, both as a response to increasing wind velocities, a lack of available soil water and a shortened growth season for reproduction. Perennial, xerophytic, evergreen plants have an advantage over deciduous species, which consume most of their stored food reserves at the commencement of their annual growth cycle. Until this latter group of plants re-accumulate carbohydrate storage products in the new season they are vulnerable to sudden changes in the climate which may affect their growth and survival.

When temperatures become favourable for growth in the spring, plants of high altitude and latitude break their dormancy, commence vegetative growth, flower and set fruit within a relatively short period which may only be of a few weeks' duration. Many herbaceous perennials have shoot and flower buds preformed one or two seasons previously, which enables them to achieve shoot elongation and flowering immediately after snow-melt. In order to protect these buds over winter some plants have clusters of protective twigs or dead leaves surrounding the vulnerable organs.

On the climatically severe, exposed, windswept ridges and late-lying snow beds, vascular plants cannot survive and the principal species are lichens on dry rock and soil and bryophytes where the substrate or atmosphere is moister. Lichens, although slow-growing, are long-lived and may remain dormant for extended periods during unfavourable conditions. They take advantage of the higher temperatures of the rock and soil surfaces on which they grow and where they can survive long periods of desiccation. Similarly, several species of bryophytes are resistant to the harsh conditions imposed by low temperatures and long periods without water.

9.2.2 *Life cycles*

Flowering plants have evolved several strategies to overcome problems of competition for pollinators and a shortened life-cycle under the conditions of environmental stress imposed by high altitudes and latitudes.

a

b

9.4 Viviparous plants: (a) *Festuca vivipara*; (b) *Polygonum viviparum*, with bulbils replacing flowers on the lower part of the flowering spike.

Firstly, since insects are less numerous, the flowers of many plants have large, open, brightly coloured and strongly scented flowers, which increase their chances of pollination and seed production. Many of the most attractive alpine flowers are in this category. Self-fertilisation and apogamy (reproduction without fertilisation) are also common. Secondly, wind-pollinated plants (e.g. grasses, rushes and sedges) increase in abundance. Thirdly, methods of vegetative reproduction are favoured and certain species adopt the viviparous mode of reproduction (Fig. 9.4) producing small plantlets (bulbils) in place of floral structures. Vivipary is exhibited by *Poa alpina* (alpine meadow-grass), *Festuca vivipara* (viviparous fescue), *Polygonum viviparum* (alpine bistort) and *Saxifraga cernua* (drooping saxifrage).

9.2.3 *Physiological adaptations*

The physiology of the whole plant determines the ability of a species to survive in northern or mountain environments. Physiological adaptations within plants include alteration of the rates of respiration and photosynthesis and an ability to metabolise successfully at low temperatures. Plants may begin their annual growth cycle as early as April or as late as September, but however late, the short growing season must be exploited to the full. Carbohydrate reserves are rapidly used up in the formation of new shoots and leaves, although new storage soon begins again as a result of intensive photosynthesis.

High levels of ultraviolet light are potentially harmful to plants and many species found growing at high altitudes have a thick, waxy epidermis and intracellular anthocyanin or flavonoid pigments in their leaves, which may help to protect them from damage. The anthocyanins, together with the green of chlorophyll, result in a dark purple to black colouration in the shoots of many plants.

9.2.3.1 *Dormancy and overwintering*

Dormancy is important in ensuring the survival of many plant species during the coldest season and yet, compared to temperate perennials, dormancy is less common in polar and mountain plants. This apparent anomaly arises because plants of high altitudes and latitudes are capable of resuming growth immediately environmental conditions become suitable, no matter how early or late in the year that may be. In particular, they possess buds which can detect small variations in temperature and respond quickly to the commencement of the growing season. Consequently, overwintering buds of arctic perennials, unlike those of temperate species, do not have thick insulating scales. Plants of extremely cold environments are also frost-resistant. Overwintering tissues, after a period of hardening, can undergo supercooling (down to −38°C). High concentrations of soluble carbohydrates in the plant cells (e.g. raffinose) may act as a natural antifreeze, preventing the cell sap from freezing, and any ice that forms is extracellular, so that the chances of tissue damage are diminished. (Fitter and Hay, 1981).

9.2.3.2 *Photosynthesis and respiration*

Plants of high altitudes and latitudes maximise the efficiency of their metabolic processes during the relatively short part of the year when both light intensity and temperature are favourable for growth. In these species the optimum rate of photosynthesis occurs at much lower temperatures than in plants of temperate environments. The minimum temperature at which photosynthesis can take place is also lower. For example, *Oxyria digyna* (mountain sorrel) can photosynthesise at summer temperatures as low as −6°C. Productivity of arctic and alpine plants appears low when expressed as an annual increase in dry weight per unit area, but when daily rates of dry matter accumulation during the growing season are considered, these can be similar to those of temperate vegetation, despite the much lower temperatures.

9.2.3.3 *Water relations*

Drought stress to plants occurs in both northern and mountain environments largely as a result of frozen soils and strong drying winds. Tundra receives low annual precipitation (section 4.1) but even on mountains where rainfall is substantial, steep slopes and rocky soils encourage rapid run-off or sub-surface seepage of water. Xeromorphic adaptations to plant leaves and stems, especially waxy or hairy leaf coverings and inrolled leaves, protect the stomata and reduce transpiration.

9.3 Arctic and alpine elements of the British flora

Many of the rarer plants of the British Isles are presently found only in mountainous situations or in the far north of Scotland. These species have strong affinities in their geographical distribution with central European alpine or northern latitude arctic environments. Although the exact reasons for the presence of these species is unclear,

it is probable that major climatic and topographical changes brought about by successive glaciations were responsible for their incorporation into the British flora. (Matthews, 1955).

Ice advanced across the British Isles several times in the million years of the Pleistocene period and on each occasion the extent of the ice cover across the country was different. At no time, however, was the whole country (or northern France) covered by ice. The forefront of the ice during the most extensive glaciation came as far south as the Rivers Thames and Severn, leaving the extreme south of England relatively ice-free. In the last glacial period, which ended about 10,000 years ago, much of England south of a line from the Humber to the Bristol Channel remained uncovered by ice. Although the north of England and the whole of Scotland were more heavily glaciated than the south of Britain the summits of the highest mountains were never completely covered by ice and remained projecting above as 'nunataks', or bare rocky islands in a glacial sea. It is thought that these outcrops may have provided refuges for the hardiest of plants during each glaciation.

It is generally accepted in biogeography that plants and animals migrate under the stress imposed by changing climatic conditions. With the commencement of each glacial period the onset of harsh environmental conditions must, therefore, have induced major changes in regional floras throughout the British Isles. Temperate species would have been gradually replaced by cold-tolerant ones from the north as the ice sheets moved progressively southwards, preceded by periglacial conditions. In addition, because of a simultaneous drop in ocean levels, continuity of the land surface between the south of Britain and continental Europe was established, over which plants could have migrated, thereby escaping the rigours of the cold climate. Similar southwards species migrations would have occurred throughout the northern hemisphere. Consequently, many plant species from Britain and northern Eurasia migrated southwards into the relatively ice-free Franco-British region.

During these glacial periods ice sheets also advanced outwards from the mountainous areas of southern Europe. As a result, species from the European Alps may have migrated northwards to the same ice-free region in the south of Britain and north of France that was colonised by northern European species. The migration route to the north was probably the relatively ice-free periglacial corridor which extended from the Mediterranean, along the south coasts of France and

Spain, to the north of France. Therefore, species whose origins were either in the arctic zone or the alpine regions of central Europe could have occupied the same territory near to the south of Britain.

With the retreat of the ice at the end of each glaciation plant migrations would have recommenced. Temperate species replaced the arctic and alpine ones in southern Britain and northern France once more as the climate ameliorated, and the hardier species moved both northwards and southwards in the wake of the melting ice. However, since plants have no sense of direction there resulted a mixing up of elements in the British flora after each glaciation! Some former alpines migrated to the arctic regions of northern Europe and former arctic species travelled southwards to the central European Alps. The result for the British flora is the presence of species whose main centres of distribution at the present day are in either the arctic, the European Alps, or both. These make up the arctic, alpine and arctic-alpine elements of the British flora.

Evidence to support these migratory theories is contained in late Pleistocene deposits from the valleys of the Rivers Lea and Cam to the north of London (Godwin, 1975). Fossilised plant remains from these sites confirm the presence in lowland Britain during the last glaciation of species whose principal centres of distribution are now either within the Arctic Circle, in the mountains of central Europe, or both. In the British Isles these species now only grow on the higher mountains and moorlands, and are often restricted to a few localities. (Pennington, 1969).

9.3.1   *Arctic species*

This group consists of nearly 30 species whose distribution in Europe is in boreal and arctic regions (Fig. 9.5). Several of these, including *Carex saxatilis* (russet sedge), *Diapensia lapponica* (diapensia), *Luzula arcuata* (curved woodrush), *Saxifraga cespitosa* (tufted saxifrage) and *S. rivularis* (highland saxifrage) are

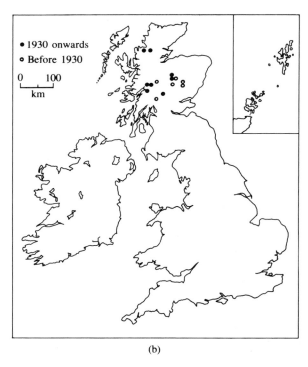

(a)

(b)

9.5 *Saxifraga rivularis* – a British arctic species: (a) world distribution (redrawn from Hultén, 1971); (b) British distribution (from Perring & Walters, 1962).

circumpolar in distribution. Others have a much more localised distribution, e.g. *Arenaria norvegica* (arctic sandwort), which, outside the British Isles, occurs only in Scandinavia and Iceland. In the British Isles most of these species are restricted to the mountains of the central Highlands of Scotland, with the exception of *Rubus chamaemorus* (cloudberry), which is widely distributed on moorlands as far south as the southern Pennines of England.

9.3.2 *Alpine species*

This component of the British flora is a small one and contains probably no more than 10 species which have their main centre of distribution in the European Alps and are absent from arctic regions (Fig. 9.6). Unlike the arctic element, many of the true alpine species have a wide distribution in the British Isles: *Gentiana verna* (spring gentian) occurs both in Teesdale and western Ireland, and *Minuartia sedoides* (cyphel) grows on unstable ground on many of the higher Scottish mountains.

9.3.3 *Arctic-alpine species*

This, the largest group of the cold environment plants in the British flora, has a dual distribution: above the tree-line in central Europe and beyond that limit in northern latitudes (Fig. 9.7). These species often have a marked preference for nutrient-enriched substrates in high mountain areas of Britain and are thus very discontinuous in their distribution. Of the more than 70 species which reputedly make up this group, nearly all occur in Scotland, with the notable exceptions of *Lloydia serotina* (Snowdon lily) in Snowdonia, and *Minuartia stricta* (Teesdale sandwort) in Teesdale.

Most of our arctic-alpine species are circumpolar, e.g. *Dryas octopetala* (mountain avens), *Oxyria digyna*, *Salix reticulata* (net-leaved willow), *Saxifraga oppositifolia* (purple saxifrage), *Silene acaulis* (moss campion) and *Thalictrum alpinum* (alpine meadow rue). Others are strict European arctic-alpines, such as *Cicerbita alpina* (alpine sowthistle); *Salix myrsinites* (whortle-leaved willow) and *S. lapponum* (downy willow) are Eurasian; whilst *Alchemilla alpina* (alpine lady's mantle), *Gentiana nivalis* (alpine gentian) and *Saxifraga stellaris* (starry saxifrage) are amphi-Atlantic (i.e. restricted to Atlantic seaboards on both sides of that ocean).

9.4 Vegetation of upland and northern Britain

The highest ground in Britain lies to the north of a line from the Bristol Channel to the Humber. South of this there is very little land over 250 m (Fig. 9.9). The distribution of high ground in Britain shows a good correlation with areas of high rainfall and the

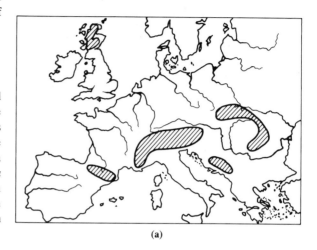

(a)

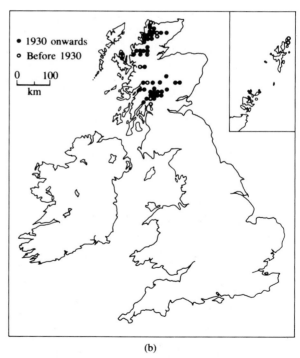

• 1930 onwards
○ Before 1930

0   100
└────┘
km

(b)

9.6 *Minuartia sedoides* – a British alpine species: (a) world distribution (redrawn from Raven & Walters, 1956); (b) British distribution (from Perring & Walters, 1962).

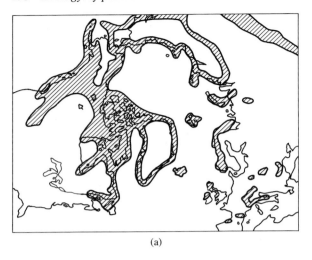

(a)

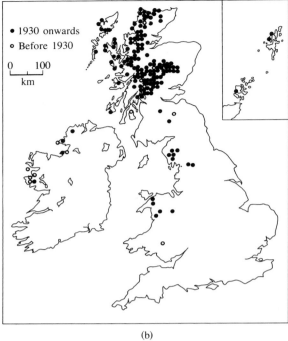

- ● 1930 onwards
- ○ Before 1930

0   100

km

(b)

9.7   *Saxifraga oppositifolia* – a British arctic-alpine species: (a) world distribution (from various sources); (b) British distribution (from Perring & Walters, 1962).

9.8   British arctic-alpine species: (a) *Saxifraga oppositifolia*; (b) *Dryas octopetala*; (c) *Oxyria digyna*.

distribution of Paleozoic rocks, which, apart from the Carboniferous limestone, give rise to nutrient-deficient, leached, acid soils (see Fig. 2.3). The principal regions with montane vegetation are in Wales, the north and south Pennines of England, the Lake District, south and west Scotland and the Scottish Highlands.

In southern Scotland and most of the upland regions of England (apart from the highest peaks in the Lake District and Snowdonia) the summits are sufficiently low to have been wooded in the past. Following deforestation many of these were converted to 'sheep walk' (i.e. rough grazing) with a consequent decrease in species diversity (section 5.3). Paradoxically, these are the areas which have now been extensively planted with conifers. The most botanically interesting areas are those where calcareous or schistose rocks enrich the soils, notably in Snowdonia, Upper Teesdale and several locations in the Lake District and the Scottish Highlands (e.g. Beinn Laoigh, Argyllshire; Ben Lawers, Perthshire; Glen Clova, Angus; and the outcrops of Durness limestone between Skye and the Pentland Firth).

9.4.1 *Vegetation zonation*

Vegetation zonation is apparent with increases in both altitude and latitude. In Britain most of the native forest has long since been removed, but where some still remains, oak, pine and birch woodland extends up the lower mountain slopes, reaching high altitudes in sheltered south-facing ravines. In general, oak grows at low altitudes, especially on richer soils and near western coasts; in Scotland, pine supersedes oak at higher latitudes and on poor, leached podzols and river gravels at low altitudes; birch replaces both oak and pine as the tree-line species at the highest altitudes and latitudes. Above the forest limit the maximum vegetation height progressively decreases as ericaceous shrubs are replaced by low-growing sedge and dwarf-willow heaths. On the highest and most exposed areas the vegetation is a poorly developed moss-lichen heath in which there is a high proportion of bare rock and soil.

Although the vegetation of upland and northern Britain includes associations from several phytosociological classes, these possess several unifying characteristics – notably the presence of arctic and alpine plants and a paucity of lowland species, which may be completely absent at the highest elevations and latitudes. There is also a high proportion of non-vascular

cryptogams (bryophytes and lichens). Bryophytes are particularly abundant in springs and flushes and also downslope from late-lying snow beds. Some mosses and many lichens dominate drier, more exposed situations, especially on the skeletal soils and rocks of windswept ridges and summits.

Further information on British montane vegetation is available in McVean and Ratcliffe (1962), Pearsall (1971) and Birks (1973).

9.4.2 *Montane grasslands and grass-heaths (Nardo-Callunetea)*

Sub-montane grasslands and heaths of the Nardo-Callunetea are widely distributed throughout the uplands of England, Wales, south and south-west Scotland, and on the lower slopes of the higher Scottish mountains,

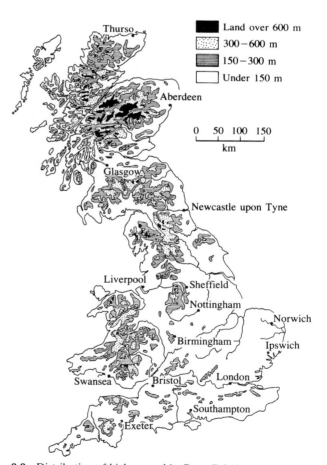

9.9 Distribution of high ground in Great Britain.

where they are characterised by the almost exclusive presence of lowland and middle altitude species (sections 5.3.1 and 5.4.1). However, with increase in altitude and latitude, arctic and alpine plants appear in increasing number in the vegetation.

9.4.2.1 *Acid grasslands (Nardetalia)*

There are several types of acid grassland which have developed in response to variations in climate, physiography, length of snow cover, soil moisture, slope, aspect and wind exposure. They are most conveniently considered collectively within the alliance Nardeto-Caricion bigelowii.

Grass-heaths dominated by *Nardus stricta* (mat-grass) occur up to 1300 m. With increase in altitude or latitude the abundance of bryophytes increases and although *Racomitrium lanuginosum* (woolly hair-moss) predominates, other mosses and liverworts more commonly associated with low altitude forests also occur (section 5.2.2.1), e.g. the feather-mosses *Hylocomium splendens*, *Pleurozium schreberi* and *Rhytidiadelphus loreus*, and the liverwort *Ptilidium ciliare*. At the highest altitudes in the British Isles where *Nardus stricta* is still present, the cryptogam layer is often dominated by the mosses *Dicranum fuscescens* and *D. scoparium* and many lichens, particularly of the genera *Cetraria* and *Cladonia*. Where drainage is poor, and the surface becomes waterlogged, there is an increase in species characteristic of ombrotrophic mires such as *Trichophorum caespitosum* (deer-grass) and *Narthecium ossifragum* (bog asphodel) (section 7.3.2).

Under the influence of intermittent irrigation water flushing over the ground the species diversity of *Nardus* grasslands increases owing to the development of soils with only moderate acidity. In these dominance is assumed by *Agrostis capillaris* (common bent-grass) and *Festuca ovina* (sheep's fescue); *Nardus stricta* is less common. Other plants frequently present are *Alchemilla alpina*, *Anthoxanthum odoratum* (sweet vernal-grass), *Galium saxatile* (heath bedstraw), *Viola riviniana* (common violet), *Thymus drucei* (wild thyme) and the mosses *Hylocomium splendens* and *Pleurozium schreberi*. These bent-fescue grasslands are of widespread occurrence in suitable habitats throughout the Scottish Highlands and on the highest mountains of the English Lake District. However, they are best developed on the mica schist soils of the Breadalbane and Clova regions of the central Highlands. They are usually heavily grazed owing to the abundance of palatable species.

9.10 Moss-lichen heath, the characteristic vegetation of summits and ridges on the higher mountains of the British Isles.

With increasing severity of climate and length of snow-lie *Carex bigelowii* (stiff sedge) assumes increasing importance in mountain vegetation, eventually replacing *Nardus stricta*. This sedge forms associations with the mosses *Polytrichum alpinum* and *Dicranum fuscescens*. In Britain, these moss-sedge heaths are confined to the central Scottish Highlands.

On exposed, windswept ridges and summits where the soils are immature, unstable rankers, *Nardus stricta* is accompanied by *Juncus trifidus* with which it forms low-growing, grassy heaths throughout the Scottish Highlands at altitudes between 700 and 1200 m. The vegetation cover ranges from a very sparse to a completely closed *Juncus trifidus-Racomitrium lanuginosum* heath in which additional low-growing species are *Salix herbacea* (dwarf willow), *Deschampsia flexuosa*, *Festuca vivipara* and *Alchemilla alpina*. Of very occasional occurrence are *Minuartia sedoides* and the rare *Diapensia lapponica* and *Artemisia norvegica*.

9.4.2.2 *Montane and sub-montane ericaceous heaths (Calluno-Ulicetalia)*

Much of the Highland zone, especially in Scotland above an altitude of 350 m in the north and west and above 700 m in the central Highlands, is dominated by shrubby members of the Ericaceae, principally *Calluna vulgaris* (heather), *Vaccinium myrtillus* (bilberry), *V. vitis-idaea* (cowberry), *Empetrum nigrum* (crowberry) and *Arctostaphylos uva-ursi* (bearberry). These heaths occur on peaty podzols and are in the alliance Empetrion nigri of the order Calluno-Ulicetalia. Additional species include *Rubus chamaemorus* (cloudberry), *Cornus suecica* (dwarf cornel) and the grasses *Nardus stricta* and *Deschampsia flexuosa* (wavy hair-grass); in places the prostrate form of *Juniperus communis* subsp. *alpina* (juniper) forms a low-growing heath, especially on level or gently sloping, immature, shallow rankers. The species of the moss layer are similar to those of low altitude northern heaths (section 5.4.1). With increase in altitude and latitude these heaths are replaced by those of the class Loiseleurio-Vaccinietea.

9.4.3 *High altitude and latitude dwarf-shrub heaths (Loiseleurio-Vaccinietea)*

The vegetation of some exposed mountain summits and ridges and of the most northerly coastal regions of Britain is tundra-like heath dominated by low-growing ericaceous shrubs, cushion-forming herbs, grasses, bryophytes and lichens. The soils on which these heaths develop are immature, unstable and liable to solifluction. The numerous associations within this class belong to the alliance Loiseleurio-Arctostaphylion. Species which occur throughout the associations of this alliance include *Loiseleuria procumbens* (trailing azalea), *Empetrum hermaphroditum* (mountain crowberry), *Vaccinium vitis-idaea* (cowberry), *Carex bigelowii*, *Deschampsia flexuosa*, *Nardus stricta*, *Agrostis canina* and *Festuca ovina*; the bryophytes *Racomitrium lanuginosum*, *Pleurozium schreberi* and *Ptilidium ciliare*; and the lichens *Cetraria islandica*, *Cladonia gracilis* and *C. uncialis*. Above 600 m in the northern Scottish Highlands *Arctostaphylos alpinus* (black bearberry) is locally common; this prostrate shrub also descends to coastal clifftops in the extreme north of Scotland.

9.4.4. *Chionophilous dwarf willow and bryophyte-dominated communities (Salicetea herbaceae)*

In the central and western Highlands of Scotland on sites over 900 m subject to moderate snow cover a low-growing vegetation occurs in which *Salix herbacea* predominates. The associations of this class are usually confined to exposed sites which receive intermittent irrigation from melting snow. They are tolerant of prolonged snow-lie (i.e. they are chionophilous) and a short growing season. The soils on which they occur are unstable rankers or rendzinas subject to solifluction. There are two distinct groups of associations: those of acid to neutral soils of the alliance Cassiopeto-Salicion herbaceae; and those of calcium-rich habitats of the Ranunculeto-Anthoxanthion.

9.4.4.1 *Dwarf willow montane heath (Cassiopeto-Salicion herbaceae)*

On some of the highest Scottish mountains on moderately acidic to neutral soils where snow lies late into the summer, *Salix herbacea* forms associations with the bryophytes *Kiaeria starkei*, *Polytrichum alpinum*, *Racomitrium lanuginosum*, *Gymnomitrium concinnatum* and *G. varians*. These occur on unstable slopes, subject to occasional landslip, and on summits and high ridges. Associated species may include *Deschampsia cespitosa* (tufted hair-grass), *Saxifraga stellaris*, *Gnaphalium supinum* (dwarf cudweed) and *Sibbaldia procumbens* (sibbaldia).

9.4.4.2 *Herb associations of calcareous soils (Ranunculeto-Anthoxanthion)*

The low-growing herb associations of this alliance occur on calcareous rocks and soils at altitudes between 700 and 1200 m in the western Scottish highlands on sites subjected to moderate snow cover, and where surface flushing takes place by water movement or by addition of rock fragments derived from weathering limestone rocks higher upslope. The vegetation is rich in species and is characterised by low-growing, prostrate herbs and shrubs, or by dwarfed forms of taller herbs more typically found at lower altitudes.

The alliance has as constants *Salix herbacea*, *Ranunculus acris* (meadow buttercup), *Agrostis capillaris*, *Selaginella selaginoides* (lesser clubmoss), *Deschampsia cespitosa*, *Alchemilla alpina*, *Sibbaldia procumbens*, *Thymus drucei*, *Minuartia sedoides* and

a

b

c

d

9.11   Plants of acid montane grasslands and grass-heaths (Nardetalia): (a) *Carex bigelowii*; (b) *Juncus trifidus*; (c) *Alchemilla alpina*; (d) *Lycopodium alpinum*.

9.12 Plant of montane and sub-montane ericaceous heaths (Calluno-Ulicetalia): *Arctostaphylos uva-ursi*.

a

9.13 Plant of high altitude and latitude dwarf-shrub heaths (Loiseleurio-Vaccinietea): *Gnaphalium supinum*.

b

9.14 (a) *Salix herbacea*, a species of high mountain areas where snow lies late into spring and early summer (Salicetea herbaceae). (b) *Sibbaldia procumbens*, a plant of the highest mountain summits and ridges in the British Isles, where snow lies late and the substrates are liable to solifluction (Cassiopeto-Salicion herbaceae).

Silene acaulis. The dominant bryophytes are *Racomitrium lanuginosum* and *Polytrichum alpinum*. This habitat also provides a niche for several rare arctic-alpine plants, e.g. *Cerastium alpinum* (alpine mouse-ear), *Carex vaginata* (sheathed sedge), *Draba norvegica* (rock whitlowgrass), *Myosotis alpestris* (alpine forget-me-not) and *Veronica alpina* (alpine speedwell). In the eastern Highlands of Scotland, *Silene acaulis* is absent and dominance is assumed by a carpet of *Alchemilla alpina* and *Sibbaldia procumbens*.

On wet cliff faces or steep, rocky ground at the foot of calcareous escarpments *Saxifraga aizoides* occurs in an association which is very rich in species and includes *Alchemilla glabra*, *Parnassia palustris* (grass of Parnassus), *Pinguicula vulgaris* (common butterwort), *Polygonum viviparum*, *Ranunculus acris*, *Saxifraga oppositifolia* and the moss *Ctenidium molluscum*. *Cystopteris montana* (mountain bladder fern) is exclusive to this association.

Also within the alliance Ranunculeto-Anthoxanthion are grassland associations dominated by *Deschampsia cespitosa* subsp. *alpina* (alpine hair-grass), which are of widespread occurrence on neutral to moderately acid podzolised soils subject to water flushing following high rainfall or snow-melt. These are either species-poor, species-rich or bryophyte-rich, depending on the nature of the underlying rock, the degree of surface-water movement, aspect and rainfall. Associated species which commonly occur are *Agrostis canina*, *A. capillaris*, *Anthoxanthum odoratum*, *Galium saxatile*, *Alchemilla alpina*, *Ranunculus acris*, *Rumex acetosa* (common sorrel) and the mosses *Hylocomium splendens*, *Polytrichum alpinum*, *Rhytidiadelphus loreus* and *R. squarrosus*. (Birks, 1973).

9.4.5   *Northern montane grasslands and grass-heaths on calcareous soils (Elyno-Seslerietea)*

The principal limestone grassland vegetation of lowland and southern Britain is in the class Festuco-Brometea (section 5.3.3). In northern and upland Britain (from Durham northwards) where the wetter, colder climate leads to the development of leached, nutrient-deficient soils, even on limestone, this is replaced by associations of the Elyno-Seslerietea, which occur infrequently on shallow rendzinas at altitudes between 500 and 1000 m in the central Scottish Highlands to near sea level on the coastal limestone of Sutherland.

Some of these grasslands and heaths are species-rich and contain several plants typical of lowland habitats, such as *Plantago lanceolata* (ribwort plantain), *Thymus drucei* and *Viola riviniana*, in addition to the montane species *Silene acaulis*, *Salix reticulata* and *Carex rupestris* (rock sedge). All the associations of this class are characterised by an abundance of *Dryas octopetala* and the exclusive presence of the pleurocarpous moss *Rhytidium rugosum*.

9.4.6   *Tall herb vegetation (Betulo-Adenostyletea)*

A major difference between the mountainous regions of the British Isles and those of central and southern Europe is the paucity in the former of luxuriant, species-rich pasture dominated by tall herb vegetation. This is largely a result of geological differences, since most of the high land in Britain is formed of hard, acidic rocks which weather slowly to form nutrient-deficient, leached, acid soils. In contrast, much of the European alpine zone contains fertile soils derived from geologically younger and softer rocks (mainly limestones and sandstones). Anthropogenic influences are also important: grazing, especially by sheep, and moorland burning, much more widely practised in Britain than in

continental Europe, impose limitations to the height and species diversity of the vegetation. Consequently, in Britain, montane tall herb vegetation is restricted to steep rocky ground or to rock ledges which are inaccessible to grazing animals, and as a result of a fragmentary distribution this vegetation is difficult to classify.

Cicerbita (= *Lactuca*) *alpina* (alpine sowthistle), which gives its name to the principal alliance, Lactucion alpinae, is very rare and is restricted to a few localities in the Scottish Highlands. Many of the other continental European indicator plants do not occur in the British Isles. Constant species which do appear include *Deschampsia cespitosa*, *Luzula sylvatica* (great woodrush), *Geum rivale* (water avens), *Sedum rosea* (roseroot) and the feather-moss *Hylocomium splendens*. A large number of other species are co-dominants in different situations, including *Geranium sylvaticum* (wood crane'sbill), *Chamaenerion angustifolium* (rosebay willowherb), *Heracleum sphondylium* (hogweed), *Trollius europaeus* (globeflower), *Alchemilla vulgaris* (lady's mantle) and *Cirsium helenioides* (melancholy thistle).

9.4.7   *Alpine and sub-alpine scree vegetation (Thlaspietea rotundifoliae)*

Scree consists of accumulations of large rocks and smaller stones overlying finer gravel and sand particles; it is commonly found below mountain ridges and cliffs at the head of valleys and corries. As a result of the mobility of the substrate, stability is only temporary, thus preventing the development of a permanent climax vegetation. Plants which can survive the rigours of this habitat often have extensive underground organs, e.g. roots or rhizomes from which re-establishment can take place following the destruction of above-ground stems and leaves. They are also frequently mat- or tussock-forming species.

The vegetation of acid mountain screes is represented throughout upland Britain by the order Androsacetalia alpinae of the class Thlaspietea rotundifoliae. The principal species are *Cryptogramma crispa* (parsley fern), *Deschampsia flexuosa*, *Festuca ovina*, *Huperzia selago* (fir clubmoss) and the small, cushion-forming bryophytes *Ditrichum zonatum* and *Grimmia donniana* in addition to *Racomitrium lanuginosum*.

At high altitudes in the north-west of Scotland unstable screes support a scantier vegetation which very

a   b

9.15 Plants of tall herb vegetation (Betulo-Adenostyletea): (a) *Cicerbita alpina*; (b) *Trollius europaeus*.

occasionally includes the rare *Koenigia islandica* (Iceland purslane) known only from the Isles of Skye and Mull.

9.4.8 *Other vegetation types of upland and northern Britain*

The vegetation described so far in this chapter can be regarded as truly montane, alpine or northerly, in that the associations develop beyond the tree-line and are influenced primarily by the severity of the climate and substrate instability. However, in more sheltered situations, especially where drainage is impeded, ombrotrophic mires of the Scheuchzerietea (section 7.3.1) and Oxycocco-Sphagnetea (section 7.3.2), and transition mires of the Parvocaricetea (section 6.5.6) may develop. The latter are often found in association with spring-heads of the Montio-Cardaminetea (section 6.5.5). Although the vegetation of these classes has been described elsewhere in this book their montane or northern variants are described below.

9.4.8.1 *High mountain and northern spring communities (Montio-Cardaminetea)*

At altitudes above 550 m, and at northern latitudes, springs and flushes of the alliance Mniobryo-Epilobion (section 6.5.5) often contain a few arctic and alpine species, e.g. *Deschampsia cespitosa* subsp. *alpina*, *Veronica serpyllifolia* subsp. *humifusa* (thyme-leaved speedwell) and *Alopecurus alpinus* (alpine foxtail). Above 850 m, on ground irrigated by late snow-melt water, the moss *Pohlia wahlenbergii* var. *glacialis* forms bright yellow-green, soft, spongy carpets in which *Deschampsia cespitosa*, *Saxifraga stellaris*, *Cerastium*

a   b

9.16 Plants of high altitude and latitude springs and flushes (Montio-Cardaminetea and Parvocaricetea): (a) *Saxifraga aizoides*; (b) *Juncus castaneus*.

cerastioides (starwort mouse-ear) and *Epilobium anagallidifolium* (alpine willowherb) also grow. The leafy liverwort *Anthelia julacea* forms an almost continuous carpet on suitable sites where snow irrigation water is available throughout most of the year. The only constant phanerogam in this association is *Deschampsia cespitosa* but the bryophytes *Scapania undulata*, *Marsupella emarginata*, *Pohlia ludwigii* and *Philonotis fontana* are common. In areas influenced by mesotrophic or eutrophic ground water *Cratoneuron commutatum* plays a major part in a tufa-forming association of the Cratoneureto-Saxifragion aizoidis which develops at calcareous spring-heads between 300 and 1000 m. The large mounds which are formed are colonised by other plants including *Saxifraga aizoides*, *Festuca rubra*, *Cardamine pratensis* and *Epilobium alsinifolium*. Apart from *Philonotis fontana* other bryophytes are poorly represented.

9.4.8.2 *Nutrient-rich flushes (Parvocaricetea)*

In upland and northern parts of Britain the order Caricetalia nigrae (section 6.5.6.1) is represented by a species-diverse vegetation of mesotrophic flushes. Exclusive species to these sites are the mosses *Homalothecium nitens* and *Cinclidum stygium*, and constant species include *Selaginella selaginoides*, *Carex nigra*, *C. panicea*, *C. pulicaris*, *Polygonum viviparum*, *Thalictrum alpinum* and the moss *Philonotis fontana*.

Calcareous flushes of the Tofieldietalia (section 6.5.6.2) characterised by the presence of *Tofieldia pusilla* (Scottish asphodel), *Eriophorum latifolium, Carex panicea* and the mosses *Campylium stellatum, Cratoneuron commutatum, Drepanocladus revolvens, Ctenidium molluscum, Scorpidium scorpioides* and *Philonotis calcarea* also occur. This vegetation is restricted to nutrient-enriched, silty muds in areas of calcareous rocks mainly in upper Teesdale and southern Cumbria in England and in the Scottish Highlands.

Occasionally at high-altitude flushes throughout the Scottish Highlands a floristically diverse community, which contains many rare British species, occurs. *Juncus biglumis* (two-flowered rush), *Saxifraga aizoides, Dryas octopetala* and *Thalictrum alpinum* are of constant occurrence, whilst rare plants include *Carex atrofusca* (scorched alpine sedge), *C. microglochin* (bristle sedge), *C. saxatilis, Equisetum variegatum, Tofieldia pusilla, Juncus castaneus* (chestnut rush) and the mosses *Amblyodon dealbatus, Catoscopium nigritum* and *Meesia uliginosa.*

Chapter 10

The urban ecosystem

Although the first cities appeared more than 5000 years ago, these early urban centres were small and surrounded by an overwhelming number of rural people. The balance of population remained roughly the same for hundreds of years until comparatively recently when, with the advent of industry and technology, there was an exodus of the population from rural areas into the rapidly expanding cities. Today's cities still occupy comparatively small areas compared to the land mass as a whole, but they contain a high proportion of the world's total population. According to United Nations' estimates, 80% of the population of the USA and 60–70% of the population of Europe live in cities (Sukopp & Werner, 1982). In developing countries urbanisation of the population is occurring so rapidly that huge cities with more than 10 million inhabitants are no longer considered as unusual. In developed countries the population of central city areas is declining as the wealthier inhabitants migrate to the more pleasant surroundings of the residential suburbs. As a result the adjacent countryside is fragmented or swallowed up in urban sprawl. At the same time, the urban infrastructure is under stress as traditional industries close down and unemployment and associated poverty of the remaining population increases. However, urban land is valuable and pressures for redevelopment are leading to considerable losses of urban green space with consequent decline in habitats for wildlife.

The city habitat, which has been brought about by the transformation of the environment to suit the desires and convenience of human beings, is the most artificial of the earth's landscapes (Schmid, 1975). It is precisely because urban areas are greatly affected by man that until recently ecological studies within the urban ecosystem have been few, and urban green spaces have been treated with caution by ecologists and conservationists alike. However, over the last two decades interest in the urban environment has escalated owing to an increased public awareness, and urban areas are no longer considered to be biological deserts. Associated with this increase in public attention has been a recognition of the necessity for wildlife conservation in cities, especially in view of the decline both in the number of individuals and of species of plants and animals in the broader countryside. This change in attitudes has been hastened by studies of the importance of urban areas for maintenance of habitat and species diversity. For example, plant species mapping in Cambridgeshire (Walters, 1970) has indicated that more species can be encountered per unit area on urban land than in equal areas of the surrounding countryside. It has also been shown that in the Greater London area 61% of the plant species of Britain have at one time been recorded (Gill & Bonnet, 1973).

The lack of knowledge of cities as living space for wildlife has initiated the urgent task of investigating the ecology of the urban environment. Cities support a unique flora and fauna, which are frequently overlooked as 'weeds' or 'pests'; yet it is the pioneer and ruderal communities that have been the subject of much of the botanical research in urban areas. Extensive surveys of

urban vegetation have been carried out in many towns and cities and the wide range of habitats which these contain has been studied in detail (Emery, 1986). However, despite this increasing interest in the urban ecosystem there is as yet no universally accepted method of classification and description that can be applied to the plant communities of cities in Britain.

10.1 The urban environment

Within cities climate, nutrient cycling, energy flow and biological composition differ from the surrounding rural land. The climatic (abiotic) changes that accompany urbanisation have been recognised and studied for longer than any changes in biota (Bornkamm *et al*, 1980; Sukopp & Werner, 1982).

In large urban areas, conditions for wildlife are rather different from those pertaining in the surrounding countryside (Sukopp *et al*, 1979). For example, the temperature is higher, often by several degrees centigrade, and although rainfall may be higher, owing to increased cloud formation promoted by air pollution, humidity is generally lower. Intensive domestic and industrial activity has inevitably led to pollution and, as a result, the abundance of some plants has been dramatically

Table 10.1 Modifications to climatic parameters in cities. These are brought about through the effects of very dense building construction compared to the surrounding rural environment (Horbert, 1979).

Climate parameters	Characteristics	In comparison to the surrounding area
Air pollution	condensation	10 times more
	gaseous pollution	5–25 times more
Solar radiation	global solar radiation	15–20% less
	UV radiation: winter	30% less
	summer	5% less
	duration of sunshine	5–15% less
Air temperature	annual mean average	0.5–1.5°C higher
	on clear days	2–6°C higher
Wind speed	annual mean average	10–20% less
	calm winds	5–20% more
Relative humidity	winter	2% less
	summer	8–10% less
Clouds	overcast	5–10% more
	fog: winter	100% more
	summer	30% more
Precipitation	total rainfall	5–10% more
	less than 5 mm rainfall daily	10% more
	snowfall	5% less

reduced; although the air we breathe is now significantly cleaner than it was 20 to 30 years ago, cities do not usually support an abundant bryophyte or lichen flora. The number of lichen species in cities provides an instant indication of the levels of air pollution since this group of plants is extremely sensitive to aerial contamination. Another factor limiting the distribution of these 'lower' plants and, in particular mosses and liverworts, is that they favour damp, humid habitats, which are uncommon in dry, built-up areas.

10.1.1   *Temperature*

Within cities there is an average rise in temperature of between 2–3°C compared to the surrounding hinterland. However, this is often exceeded, with extremes recorded in excess of 11°C. This characteristic rise of temperature produces what is known as a 'heat island'. The level of increase in temperature compared to the surrounding land depends on the size of the city – usually the larger the city the greater the increase. Several factors contribute to the rise in temperature: increased thermal capacity of building materials; air pollution that produces a 'greenhouse effect'; combustion of fossil fuels in industrial and domestic systems; and reduced cooling effects through evapotranspiration owing to a reduction in the area of vegetation cover and open water surfaces. The rise in temperature results in fewer days of frost and snow and an increase in the amount of rainfall compared to the surrounding countryside. The warmer climate of cities is reflected in the urban flora, which often contains species of more southerly distribution than occur in the surrounding countryside.

10.1.2   *Rainfall and humidity*

Both the frequency and the amount of rainfall increase in urban areas, and this can be attributed to increased cloud formation promoted by air pollution and heat stagnation. The rise in rainfall can completely alter the water economy of regions adjacent to cities by increased run-off. Although humidity is generally low in the city environment there can be periodic rises in the relative humidity, especially during winter months or on still nights. The overall fall in humidity is related to the reduced area of open water, the rapid run-off of rainfall (thus less water can evaporate) and a reduction in evapotranspiration.

10.1.3   *Wind*

A wide range of wind conditions can exist within a city. Generally, vertical constructions break up the ground surface structure which leads to a reduction of the ground-level wind velocity. However, buildings can increase wind currents by causing canyon effects which greatly increase wind speeds. If a city is situated in an area of predominantly weak wind currents, then thermal turbulence can cause strong breezes within the inner areas.

10.1.4   *Pollution*

Aerial pollution causes plants to exhibit reduced vitality, accelerated senescence, reduction in biomass and disturbs reproductive capacity. This is brought about by the effects of some of the chemical compounds expelled into the atmosphere from urban areas, of which over 30 are of essential ecological importance because they enter or interfere with biological processes. The effects of gaseous pollutants on plants include cell damage in leaves, collapse of leaf tissues, reduction in plant size, root growth and leaf pigment content, early leaf fall and death. Particulate pollutants can be divided into sedimentation dusts and airborne particles. Dusts can be toxic with the presence of substances such as lead, cadmium, zinc, copper and fluoride. Sources of dust pollution are usually processes connected with industry (e.g. rubbish disposal by incineration, cement factories, metal works).

Water contamination is the other major form of pollution within urban environments, and again it is households and industry that provide the bulk of the waste material responsible. Nearly all rivers of urban areas are used as waste disposal dumps for a variety of pollutants, ranging from industrial effluent to domestic sewage. These pollutants cause eutrophication of water and increase the sedimentation load. Water pollution can also be brought about through acid rain deposition and acidity levels as low as pH 3 have been recorded in cities. Thermal pollution of rivers also occurs and water from power station cooling towers can raise water temperatures by up to 5°C downstream from the discharge point. Raised temperatures can extend the growing season of aquatic plants by up to five weeks, making them more susceptible to the effects of pollution. As a consequence of water pollution the flora and fauna of urban lakes and rivers are characteristically poor.

10.1 View of urban dereliction in Tyne and Wear.

10.1.5 *Soils*

In the city environment soils serve as the basis for construction and are greatly affected by compaction, pollution and the addition of building materials. These anthropogenic substrates can be classified as ruderal soils that support characteristic pioneer vegetation. Average pH values of 6–8.5 are frequently encountered in urban soils owing to the presence of alkaline wastes, dusts and fertilisers. The compaction of the soil causes a drastic decrease in the amount of life it supports which, along with toxic pollutants, severely reduces the soil microflora.

10.1.6 *Human disturbance*

By the very nature of their creation, urban areas have replaced the former natural, semi-natural or agricultural landscapes that preceded them. However, within towns and cities there are intense pressures on the residual and new habitats that they contain. Trampling by frequent walking, running or playing on urban land and the unauthorised use of motor vehicles causes considerable harm to urban ecosystems, as does indiscriminate dumping of refuse and other forms of vandalism. 'Damage' also results from the overzealous pursuit of landscaping and gardening practices, especially grass cutting, which may be carried out many times every year on playing fields, parks, recreation grounds and roadside reservations and verges, to the detriment of wildlife.

10.2 Urban habitats

Within urban areas there is considerable spatial differentiation and structural diversity, which provides a broad range of substrates for plant colonisation. Cities contain many different vegetation communities, from spontaneous pioneer types to managed grassland and parkland and relict natural vegetation of former rural areas. The range of habitats within urban areas is determined by the age, size, location and previous land use of the open spaces; older cities contain a larger variety of habitats and microhabitats than new conurbations. (Kunick, 1982).

The most outstanding areas of wildlife diversity within cities are semi-natural, pre-urban habitats. In most cities small remnants of the former countryside have been trapped within their urban framework during the rapid extension of the built environment during the nineteenth and twentieth centuries. These segments of 'encapsulated countryside' include woodlands, wetlands, heaths and grasslands (parklands). The quality of the vegetation in these varies, depending on the history of management and the effects of the local urban environment (including the adjacent human communities) on them.

Urban woodlands include a wide range of different types. Some cities support large areas of forest, whilst others have only small patches of trees planted on road-side verges. Some woodlands develop spontaneously in old quarries and excavation pits, along canal banks and railway lines. In Britain this type of woodland is dominated by *Acer pseudoplatanus* (sycamore), *Betula pendula* (birch) and *Fraxinus excelsior* (ash) with some *Quercus* spp. (oak) and *Fagus sylvatica* (beech). However, the ground flora of these woods is usually less diverse than that of rural examples owing to a lack of management, vandalism, rubbish tipping, trampling and pollution.

Wetlands are rare and diminishing habitats in the urban environment, where most have long since been destroyed by river canalisation or land drainage. Originally wetlands were removed for public health reasons, but as the demand for urban building space increased their decline continued. At the same time, however, new urban wetlands have been created following excavations for ballast, clay or gravel. Although restricted in distribution, urban wetlands can be very diverse and they are important biological and educational resources. In species composition they closely resemble their rural counterparts (Ch. 6) and may contain all stages of the hydroseral succession, from open-water communities through various types of reed swamp to acid bog with *Sphagnum* spp. and other acidophilous plants.

The ruderal weed communities of disturbed land are the only truly spontaneous urban vegetation types. Disturbed habitats are often associated with urban decline. They include demolition sites and derelict land, old railway lines and canals. Yet surprisingly these areas often support an abundant wildlife, including native and alien flowering plants. The ruderal plants of waste places produce abundant seed, which makes them particularly successful in the urban environment where newly exposed land is constantly available for colonisation. Railway marshalling yards, embankments and tracks (both used and disused) provide some of the most interesting disturbed habitats (Fig. 10.2). On derelict areas which have been abandoned for a long time the pioneer weed communities are replaced by taller grasses and herbs and eventually scrub invades.

Semi-natural grasslands rarely persist in the built

10.2 Abandoned railway lines and sidings are a valuable wildlife reservoir and form important corridor links between inner city areas and the surrounding countryside. Collin Street Viaduct in the centre of Nottingham is now the world's first nature reserve in the sky. Access to this site can only be gained by ladder or hydraulic platform.

environment, since this land attracts a high price for urban development. However, some small areas of acid, neutral and limestone grassland have been overlooked or neglected in the pace of urbanisation and some relict meadows remain along river sides or as part of formal parklands in areas that are less accessible to grass-cutting machines. Most of the grassland within cities is 'amenity grassland' of parks, playing fields and recreation grounds. These are usually intensively managed and often contain ornamental trees, shrubs and herbaceous species. Although they have a low species diversity they are important to local communities and often provide the only open green spaces within the built environment. With sympathetic planning and planting they can be made more attractive to wildlife. Older cemeteries also contain grassland and herbaceous species of considerable diversity.

10.3 Urban grasslands: (a) Bulwell Meadow, Nottingham, is a species-rich grassland on Magnesian Limestone, surrounded by football pitches. (b) Wilford Power Station development site, Nottingham. Dense colonies of *Dactylorhiza fuchsii* (common spotted orchid) and *D. majalis* (marsh orchid) have colonised a former pulverised fuel ash settlement pond.

Perhaps the most important and underestimated oasis for wildlife in towns and cities is the myriad of private gardens which are of infinite variety and size, from well-manicured stately lawns to disorderly jungles. Some large gardens and allotments are miniature naturescapes and many contain ponds fringed with submerged, floating and emergent water plants and abundant animal life. However, smaller garden plots are also of great importance in the urban ecosystem, since collectively they add up to a considerable area and provide the nucleus of plants for dispersal and colonisation of newly disturbed land. The true value of gardens and allotments to wildlife in cities is as yet inadequately understood and requires systematic survey and evaluation.

Green open spaces are not the only habitats of urban areas. Artificially created habitats occur in the form of walls and pavings, reproducing a similar substrate to that of cliff and rock face. As a consequence some plant species have increased their range of distribution in a dramatic fashion, e.g. *Cymbalaria muralis* (ivy-leaved toadflax) has spread from its natural habitat of rock faces and cliffs in the mountains of the southern Alps and Balkan peninsula to most parts of Britain, where it occurs extensively on walls and buildings. Roof-tops provide a habitat for crustaceous lichens. Asbestos roofs are particularly favoured and may become covered in a mosaic of coloured patches. The most frequent species are *Xanthoria parietina* which is bright orange when growing in full sunlight but yellow or green in the shade, and the mustard-yellow *Candelariella vitellina*.

Urban vegetation provides many advantages to the city environment. Apart from the aesthetic value of plants in the dull, built-up surroundings of a city, they also bring about local climatic amelioration by providing shade, reducing heat losses from the ground and absorbing aerial and aquatic pollutants. Vegetated areas can also increase the quality of life for the inhabitants of cities through noise reduction and as a facility for recreational and educational activities. In certain urban areas, particularly inner cities, there are very few wild refuges for plants and animals. Wasteland and under-used sites can be landscaped into attractive and wildlife-rich nature parks by planting new woodlands and hedgerows, creating ponds and marshes, sowing wildflower seeds to establish colourful meadows and by carefully selecting those flowers and shrubs which provide food and shelter for wildlife. Towns and cities are becoming increasingly important in terms of the total wildlife resource and many urban areas now contain a greater variety of habitats than the countryside surrounding them. This resource could be further expanded by appropriate habitat improvement and habitat creation; areas that would benefit include public parks and private gardens, cemeteries, roadside verges, disused railway lines, the grounds of public buildings, new industrial sites and derelict land.

10.4   Urban wetland: Moorbridge Pond, Nottingham, is bounded by two major roads, a railway line and a campsite.

10.3 Urban plants

Pioneer ruderal species are successful in the urban environment for a variety of reasons: they have a high reproductive potential and genetic diversity; they exhibit physiological and morphological plasticity; and they have effective seed dispersal mechanisms enabling rapid propagation. The genetic plasticity of *Oenothera* spp. (evening primrose), common plants of derelict ground, has been demonstrated since the introduction of the genus into Europe from North America. Europe now supports 13 species that differ from their North American antecedents.

Pioneer and ruderal weed communities contain a large percentage of introduced (alien) species, most of which orginate from warmer parts of the world. In some cases aliens were introduced purposely by man because of their economic importance and have now become firmly established in this country. A large number of aliens, however, have been introduced unwittingly as contaminants of imported grain, foodstuffs and raw materials such as wool from South America and Australia. Some urban alien plants have been in the British Isles for so long that they are now regarded as part of our indigenous flora. Others are of much more recent appearance and are still regarded as foreigners although they have spread to many parts of the country. One of our longest established alien plants is *Senecio squalidus* (Oxford ragwort) which was introduced into Oxford Botanic Garden in the late seventeenth century. By the mid-eighteenth century it had escaped and become naturalised in the City of Oxford, from where it has since spread throughout most of England and parts of Ireland, Scotland and Wales. The dispersal of this species is reputed to have been aided by the construction of the railway network, since the ballast used in permanent way construction provides a light, dry, stoney substrate similar to the volcanic ash soils of its native Sicily. A more recent invader is *Matricaria matricarioides* (pineapple-weed), which within 25 years of its introduction from north-eastern Asia at the end of the nineteenth century had spread throughout the country.

Urban weed floras of Britain and northern Europe include a substantial and increasing number of North American species, e.g. *Aster novi-belgii* (michaelmas daisy), *Conyza canadensis* (Canadian fleabane), *Solidago canadensis* (Canadian golden rod) and *Oenothora* spp. Other established aliens originate from Asia, e.g. *Buddleja davidii* (butterfly bush), *Reynoutria japonica* (Japanese knotweed) and *Impatiens glandulifera* (Indian balsam). However, by far the most have come from the Mediterranean region, e.g. *Hordeum murinum* (wall barley), *Cymbalaria muralis, Sisymbrium* spp. (mustard species), *Senecio squalidus* and *Diplotaxis muralis* (annual wall rocket). The commoner native weed plants include *Artemisia vulgaris* (mugwort), *Poa annua* (annual meadow-grass) and *Urtica dioica* (common nettle).

10.5 Gravestone covered in lichens. Cemeteries can provide a haven for wildlife in cities although the lichen flora is limited because of air pollution.

10.4 Vegetation of the urban environment

Different terms are used to describe the groups of plant species that grow in urban areas. There is a basic division between: naturally occurring native species (apophytes); species introduced by man intentionally and then managed; and species introduced by man but which are

a

c

b

10.6 Plants of relatively stable, formerly disturbed ground (Artemisietea vulgaris): (a) *Oenothera biennis*; (b) *Dipsacus sylvestris*; (c) *Linaria vulgaris*.

allowed to grow wild unintentionally (anthropophytes). Introduced species are further classified into those that colonised before the end of the middle ages (archaeophytes) and those species that colonised after the start of the fifteenth century (neophytes). Studies of the distribution of plant species within urban areas has confirmed that human disturbance has led to an increase in ruderal weeds and introduced species and a decrease in native wild plants. In Berlin a study of the plant species of wastelands, woodlands, dry grasslands, marshlands and aquatic habitats showed that the highest percentage of native woodland species occurred in the outer suburbs and that the highest levels of anthropophytes were found in the inner, built-up and industrial zones of the city. This has been confirmed by subsequent studies of urban vegetation, in which the largest decline of species, compared with the surrounding countryside, is in the vegetation of fields, bogs, water and rough meadow (Sukopp *et al.*, 1979; Kunick, 1982).

Urban communities of pioneer and ruderal vegetation that dominate some of the most artificial of urban habitats form consistent and separate plant associations (Haigh, 1980). Communities of the orders Plantaginetalia, Artemisietalia and Sisymbrietalia are particularly common urban vegetation types and the association Tanaceto-Artemisietum is considered to be the most completely urbanised. Only those plant associations of disturbed and derelict land, the plant species of which have colonised without assistance from man and which have developed and stabilised without management, are described here. These classes are all recognised units in the phytosociological literature. Anthropogenic variants of other vegetation types, woodlands, wetlands and grasslands, are not considered. However, their semi-natural counterparts can be found elsewhere in this book.

10.4.1 *Pioneer vegetation of open, disturbed substrates (Plantaginetea majoris)*

The associations of this class are transient successional stages produced in response to specific environmental stresses, e.g. trampling or impoverished habitat conditions. Within the single order Plantaginetalia majoris are the two alliances Lolio-Plantaginion and Agropyro-Rumicion crispi.

The former alliance contains plant communities of trampled areas consisting mainly of low-growing, decumbent or rosette-forming plants. The characteristic species are *Plantago major* (greater plantain), *Lolium perenne* (perennial rye-grass), *Taraxacum officinale* (dandelion), *Poa annua* (annual meadow-grass), *Capsella bursa-pastoris* (shepherd's purse), *Polygonum aviculare* (knotgrass), *Tussilago farfara* (coltsfoot) and the mosses *Bryum bicolor*, *B. caespiticium* and *Funaria hygrometrica*. This vegetation occurs on car parks, footpaths and playgrounds. Some species also form pioneer associations on open, calcareous, disturbed substrates.

The associations of the alliance Agropyro-Rumicion crispi embrace the range of vegetation types which develop on disturbed ground and, in particular, the zonation from wet to dry and from nutrient-rich to nutrient-poor. Some of the associations are natural, others are completely anthropogenic. The characteristic species include *Potentilla anserina* (silverweed), *Ranunculus repens* (creeping buttercup), *Leontodon autumnalis* (autumn hawkbit), *Carex hirta* (hairy sedge), *Trifolium hybridum* (alsike clover), *Elymus repens* (common couch), *Rumex crispus* (curled dock), *Pulicaria*

dysenterica (common fleabane) and *Lotus corniculatus* (common bird's-foot trefoil). Many of the species are aggressive invaders, which replace the pioneer ruderal vegetation of the Chenopodietea (section 10.4.3). They form the 'rough grasslands' of former industrial sites which have been cleared prior to redevelopment.

10.4.2 *Vegetation of relatively stable, formerly disturbed ground (Artemisietea vulgaris)*

The vegetation of this class is typical of stabilised soils in areas which in the past were severely disturbed, e.g. former industrial or housing estates. Many urban areas devastated by bombing in the last war became colonised by plant communities within this group. The variety of associations mirrors the discontinuity in soil types and the degree to which soil development and maturation has proceeded since disturbance. In general the substrates are very immature with high proportions of waste substances left over from industrial and other urban uses. However, some of these soils are rich in humus and nitrogen. The character species of the class are *Urtica dioica*, *Rumex obtusifolius* (broad-leaved dock), *Carduus acanthoides* (welted thistle), *Dipsacus sylvestris* (teasel) and *Galium aparine* (goose grass). Within the single order Artemisietalia vulgaris are the three alliances Arction, Galio- Alliarion and Aegopodion podagrariae.

The alliance Arction, which contains anthropogenic associations of stabilised nitrate- and humus-rich soils in towns and industrial areas, is characterised by the presence of either or both *Arctium lappa* (greater burdock) and *A. minus* (lesser burdock). The completely anthropogenic association Tanaceto-Artemisietum comes within this group with its diverse flora of urban ruderal plants including *Tanacetum vulgare* (tansy), *Artemisia vulgaris* (mugwort), *Solidago canadensis* (Canadian golden rod), *Linaria vulgaris* (common toadflax), *Hypericum perforatum* (perforate St John's wort), *Diplotaxis spp.* (wall rocket), *Bryonia cretica* subsp. *dioica* (white bryony) and *Oenothera biennis* (common evening primrose).

The alliances Galio-Alliarion and Aegopodion podagrariae contain few associations and are typical of garden hedges and waysides on the urban fringe, mainly in shaded situations. Characteristic species of the former alliance are *Rumex obtusifolius*, *Torilis japonica* (upright hedge parsley), *Alliaria petiolata* (garlic mustard), *Chaerophyllum temulentum* (rough chervil), *Bromus sterilis* (barren brome) and *Anthriscus sylvestris* (cow parsley). The latter alliance contains communities reminiscent of shaded woodland. These marginal zones are small and dominated by nettles and grasses. The principal species are *Aegopodium podagraria* (ground elder), *Urtica dioica*, *Lamium album* (white dead-nettle), *Elymus repens* and *Arrhenatherum elatius* (false oat-grass).

10.4.3   *Pioneer vegetation of nitrogen-rich substrates (Chenopodietea)*

The associations of this class are nitrophilous communities of arable fields and waste places. In cities the species are invasive and they occupy a wide range of habitats including footpaths, sports-fields, industrial land, spoil tips, camp sites and car parks. They occur throughout the entire infrastructure of towns and cities. The phytosociology of this group is complicated by the fragmentary nature and often small size of the habitats, differences in species available for colonisation between geographical regions and the combinations of species that

result. The character species of the class include *Chenopodium album* (fat hen), *Stellaria media* (common chickweed), *Geranium pusillum* (small-flowered cranesbill), *Solanum nigrum* (black nightshade), *Senecio vulgaris* (groundsel), *Sonchus oleraceus* (smooth sowthistle), *S. asper* (prickly sowthistle) and *Capsella bursa-pastoris*.

The main differences within the class are between the associations of the order Polygono-Chenopodietalia of nutrient-poor, dry (often sandy) substrates of marginal verges and river banks, and those of the Sisymbrietalia, which are more ruderal in complexion. The latter prefer soils with a higher humus and nitrogen content. The characteristic species of the Polygono-Chenopodietalia include *Polygonum persicaria* (redshank), *Geranium dissectum* (cut-leaved cranesbill), *Euphorbia helioscopia* (sun spurge), *Veronica agrestis* (green field speedwell), *Lamium purpureum* (red dead-nettle), *Stachys arvensis* (field woundwort) and *Thlaspi arvense* (field penny cress). The Sisymbrietalia contains a large number of species in several associations, including *Malva sylvestris*

(a)   (b)

10.7   (a) *Agrimonia eupatoria*, a plant of hedgebanks, roadsides and dry grassland. (b) *Chamaenerion angustifolium*, a nitrophilous species of disturbed land (Epilobietea angustifolii).

(common mallow), *Diplotaxis* spp., *Lactuca serriola* (prickly lettuce), *Conyza canadensis, Sisymbrium officinale* (hedge mustard), *S. altissimum* (tall rocket), *Vulpia myuros* (rat's-tail fescue) and *Lepidium campestre* (field pepperwort).

10.4.4 *Nitrophilous communities of disturbed sites (Epilobietea angustifolii)*

This vegetation forms fragmentary stands at the edge of other classes. These are species-poor and dominated by tall annuals, biennials and low-growing perennials, especially on humic, mineral-rich soils. The associations predominate on disturbed ground where the taller vegetation has been removed. Characteristic species are *Chamaenerion* (=*Epilobium*) *angustifolium, Urtica dioica* and *Rubus fruticosus* agg. (bramble). In urban areas the associations of the Epilobietea angustifolii within the order Epilobietalia angustifolii and alliance Epilobion angustifolii are very fragmentary and their phytosociology requires detailed investigation. The species overlap with those of the Sambucetalia of the class Rhamno-Prunetea, which also develops fragmentary associations in towns and cities and it may be that the two should be merged within one class.

10.4.5 *Fern-dominated communities of rocks and walls (Asplenietea rupestris)*

The completely artificial habitats of buildings and paved areas, crevices in structures and brickwork and small, shady areas in roadways, cuttings and embankments, provide the niches for cryptogam-dominated communities whose habitats resemble those of cliff- and rock-face elsewhere in the countryside. The plants and pioneer species and the resultant vegetation is sparse owing to the severity of the microclimate which extends over extremes of light, temperature, water availability and exposure. The substrates range from acid to alkaline and the soils are primitive and poorly developed, resembling rocky rankers on the one hand and immature rendzinas on the other.

Within the single order Tortulo-Cymbalarietalia are the two alliances Parietarion judaicae and Cymbalario-Asplenion. Associations of the former contain species with a southerly distribution in Europe and which prefer warm habitats. Of the characteristic species the moss *Tortula muralis* grows in tufts and cushions on mortar, concrete and basic rocks and walls, usually in exposed situations, whilst *Parietaria judaica* (pellitory of the wall) and *Cheiranthus cheiri* (wallflower) have a preference for warm, sheltered locations on walls and buildings where organic debris has accumulated, but also on calcareous substrates such as cement and mortar.

Associations of the Cymbalario-Asplenion are more diverse in species composition than the Parietarion judaicae and contain pioneer species of walls and buildings, especially in cool, damp, north-facing situations. Characteristic species include the ferns *Asplenium ruta-muraria* (wall rue), *A. trichomanes* (maidenhair spleenwort), *Polypodium vulgare* (polypody) and *Asplenium scolopendrium* (hart's tongue fern); the bryophytes *Rhynchostegium murale, Barbula unguiculata, Encalypta streptocarpa, Homalothecium sericeum* and *Marchantia polymorpha*; and the flowering plants *Cymbalaria muralis, Tanacetum parthenium* (feverfew), *Epilobium montanum* (broad-leaved willow-herb), *Sagina procumbens* (procumbent pearlwort), *Dactylis glomerata* (cocksfoot) and *Poa annua*.

References

Ahmad-Shah, A. A. (1984) Plant nutrient fluxes in an afforested mire, Ph.D. Thesis, University of Nottingham.

Allen, S. E. (1964) Chemical aspects of heather burning, *J. appl. Ecol.* **1**, 347–67.

Allen, S. E., Carlisle, A., White, E. J. & Evans, C. C. (1968). The plant nutrient content of rainwater, *J. Ecol., 56*, 497–564.

Anderson, J. A. R. (1964) The structure and development of the peat swamps of Sarawak and Brunei, *J. Trop. Geogr.* **18**, 7–16.

Armstrong, W. (1967) The oxidising activity of roots in waterlogged soils, *Physiol, Plant.* **20**, 920–6.

Bär, O. (1976) *Geographie der Schweiz.* Lehrmittelverlag des Kantons, Zurich.

Barbour, M. G., Burk, J. H. & Pitts, W. D. (1987) *Terrestrial Plant Ecology*, 2nd edn. Benjamin/Cummings, California.

Bartley, D. D. (1960) Rhosgoch Common, Radnorshire: stratigraphy and pollen analysis, *New Phytol.* **59**, 238–62.

Becking, R. W. (1957) The Zurich-Montpellier school of phytosociology, *Bot. Rev.* **23**, 411–88.

Begon, M., Harper, J. C. & Townsend, C. R. (1986) *Ecology: individuals, populations and communities.* Blackwell, Oxford.

Birks, H. J. B. (1973) *Past and Present Vegetation of the Isle of Skye: a palaeoecological study.* Cambridge University Press, London.

Birks, H. J. B., Deacon, J. & Peglar, S. (1975) Pollen maps for the British Isles 5000 years ago, *Proc. Roy. Soc. Lond. B*, **189**, 87–105.

Boatman, D. J. (1983) The Silver Flowe National Nature Reserve, Galloway, Scotland, *J. Biogeog.* **10**, 163–274.

Boatman, D. J., Hulme, P. D. & Tomlinson, R. W. (1975) Monthly determinations of the concentrations of sodium, potassium, magnesium and calcium in the rain and pools on the Silver Flowe National Nature Reserve, *J. Ecol.*, **63**, 903–12.

Bornkamm, R., Lee, J. & Seaward, M. (eds) (1980) *Urban Ecology.* Blackwell Scientific Publications, Oxford.

Bourgeron, P. S. & Guillaumet, J. L. (1982) Vertical structure of trees in the Tai forest (Ivory Coast): a morphological and structural approach, *Candollea* **37**, 565–77.

Bower, M. M. (1962) The causes of erosion in blanket peat bogs, *Scottish Geographical Magazine* **78**, 33–43.

Brady, N. C. (1974) *The Nature and Properties of soils.*, 8th edn. Macmillian, New York & Collier Macmillian, London.

Braun-Blanquet, J. (1928) *Pflanzensoziologie. Grundzuge der Vegetationskunde.* Springer, Berlin. Translated (1932) *Plant Sociology; the study of plant communities.* McGraw-Hill, New York.

Braun-Blanquet, J. & Tüxen, R. (1952) Irische Pflanzengesellschaften, *Inst. Rubel Zurich* **25**, 224–421.

Brown, D. H. (1982) Mineral Nutrition, Ch. 11, pp. 383–444, in Smith, A. J. E. (ed.) *Bryophyte Ecology.* Chapman and Hall, London & New York.

Carlisle, A., Brown, A. H. F. & White, E. J. (1966) The organic matter and nutrient elements in the precipitation beneath sessile oak canopy, *J. Ecol.*, **54**, 87–98.

Chapman, V. J. (1964) *Coastal Vegetation.* Pergamon Press, Oxford. (2nd edn 1976.)

Clapham, A. R., Tutin, T. G. & Moore, D. M. (1987) *Flora of the British Isles*, 3rd edn. Cambridge University Press, Cambridge.

Clymo, R. S. (1963) Ion exchange in *Sphagnum* and its relation to bog ecology, *Ann. Bot. (Lond.) N.S.* **27**, 309–24.

Clymo, R. S. (1967) Control of cation concentration, and in particular of pH, in *Sphagnum*-dominated communities, pp. 273–84 in Golterman, H. L. & Clymo, R. S. (eds) *Chemical Environment in the Aquatic Habitat.* North Holland, Amsterdam.

Clymo, R. S. & Hayward, P. M. (1982) The ecology of *Sphagnum*, Ch. 8, pp. 229–89 in Smith, A. J. E. (ed.) *Bryophyte Ecology.* Chapman and Hall, London & New York.

Coles, J. M. & Hibbert, F. A. (1968) Prehistoric roads and tracks in Somerset, England. 1. Neolithic, *Proc. prehist. Soc.* **34**, 238.

Coles, J. M., Hibbert, F. A. & Clements, F. C. (1970) Prehistoric roads and tracks in Somerset, England. 2. Neolithic, *Proc. prehist. Soc* **36**, 125.

Corley, M. F. V. & Hill, M. O. (1981) *Distribution of Bryophytes in the British Isles.* British Bryological Society, Cardiff.

Crawford, R. M. M. (1972) Some metabolic aspects of ecology, *Trans. bot. Soc. Edinb.* **41**, 309–22.

Crawford, R. M. M. (1978) Metabolic adaptations to anoxia, pp. 119–36 in Hook, D. D. & Crawford, R. M. M. (eds) *Plant Life in Anaerobic Environments.* Ann Arbor Science Publishers, Michigan.

Crawford, R. M. M. (1983) Root survival in flooded soils, Ch. 7, pp. 257–83 in Gore, A. J. P. (ed.) *Ecosystems of the World 4A. Mires: Swamp, Bog, Fen and Moor.* Elsevier, Amsterdam.

Crisp, D. T. (1966) Input and output of minerals for an area of Pennine moorland. The importance of precipitation, drainage, peat erosion and animals, *J. appl. Ecol.* **3**, 327–48.

Cruise, J. & Newman, A. (1973) *Photographic Techniques in Scientific Research*, vol. 1. Academic Press, London.

Curtis, L. F., Courtney, F. M. & Trudgill, S. T. (1976) *Soils in the British Isles.* Longman, London and New York.

Dajoz, R. (1977) *Introduction to Ecology.* Hodder and Stoughton, London.

Darby, H. C. (1951) The clearing of the English woodlands, *Geography* **36**, 71–83.

Davies, J. N., Briarty, L. G. & Rieley, J. O. (1972) Observations on the swollen lateral roots of the Cyperaceae, *New Phytol.* **72**, 167–74.

Dennington, V. N. & Chadwick, M. J. (1978) The nutrient budget of colliery spoil tip sites. I. Nutrient input in rainfall and nutrient losses in surface run-off, *J. appl. Ecol.*, **15**, 303–316.

Dickinson, C. H. (1983) Micro-organisms in peatlands, Ch. 5, pp. 225–45 in Gore, A. J. P. (ed) *Ecosystems of the World 4A. Mires: Swamp, Bog, Fen and Moor.* Elsevier, Amsterdam.

Dimbleby, G. W. (1962) *The Development of British Heathlands and their Soils.* Clarendon, Oxford.

Dony, J. G., Perring, F. H. & Rob, C. M. (1980) *English Names of Wild Flowers.* Butterworths, London.

Ellenberg, H. (1963) *Vegetation Mitteleuropas mit den Alpen.* Eugen Ulmer, Stuttgart.

Emery, M. (1986) *Promoting Nature in Cities and Towns.* Croom Helm, London.

Epstein, E. (1972) *Mineral Nutrition of Plants: Principles and Perspectives.* Wiley, New York.

Etherington, J. R. (1975) *Environment and Plant Ecology.* John Wiley, London. (2nd edn 1982.)

Eyre, S. R. (1968) *Vegetation and Soils*, 2nd edn. Edward Arnold, London.

Ferguson, N. P., Lee, J. A. & Bell, J. N. B. (1978) Effects of sulphur pollutants on the growth of *Sphagnum* species, *Environ. Pollut.* **16**, 151–61.

Firbas, F. (1952) Einige Berechnungen über die Ernährung der Hochmoore, *Veroff. geobot. Inst. Rubel, Zurich* **25**, 177–200.

Fitter, A. H. and Hay, R. K. M. (1981) *Environmental Physiology of Plants.* Academic Press, London. (2nd edn 1987).

Flowers, T. J., Troke, P. F. & Yeo, A. R. (1977). The mechanism of salt tolerance in halophytes, *Ann. Rev. Plant Physiol.* **28**, 89–121.

Gill, D. & Bonnet, P. (1973) *Nature in the Urban Landscape: a Study of City Ecosystems.* York Press, Baltimore.

Gimingham, C. H. (1972) *Ecology of Heathlands.* Chapman and Hall, London.

Godwin, H. (1962) Vegetational history of the Kentish chalk downs as seen at Wingham and Frogholt, *Veroff. geobot. Inst. Zurich* **37**, 83–99.

Godwin, H. (1975) *The History of the British Flora: a factual basis for phytogeography*, 2nd edn. Cambridge University Press, Cambridge.

Goodman, G. T. & Perkins, D. F. (1959) Mineral uptake and retention in cotton-grass (*Eriophorum vaginatum* L.), *Nature, Lond.* **184**, 467–8.

Gore, A. J. P. (1968) The supply of six elements by rain to an upland peat area, *J. Ecol.*, **56**, 483–95.

Gore, A. J. P. (ed.) (1983) Introduction, Ch. 1, pp. 1–34, *Ecosystems of the World 4A. Mires: Swamp, Bog, Fen and Moor.* Elsevier, Amsterdam.

Green, B. H. (1968) Factors influencing the spatial and temporal distribution of *Sphagnum imbricatum* Hornsch. ex Russ. in the British Isles, *J. Ecol.* **56**, 47–58.

Green, B. H. & Pearson, M. C. (1977) The ecology of Wybunbury Moss, Cheshire. I. The present vegetation and some physical, chemical and historical factors controlling its nature and distribution, *J. Ecol.* **65**, 793–814.

Haigh, M. J. (1980) Ruderal communities in English cities, *Urban Ecology* **4**, 329–38.

Harper, J. L. (1977) *The Population Biology of Plants.* Academic Press, London.

Haslam, S. M. (1978) *River Plants.* Cambridge University Press, Cambridge.

Horbert, M. (1979) Klimatische und lufthygienische Aspekte der Stadt- und Landschaftsplanung, *Natur und Heimat* **38**, 34–49.

Hoskins, W. G. (1955) *The Making of the English Landscape.* Hodder & Stoughton, London.

Hulten, E. (1971) *The Circumpolar Plants. II. Dicotyledons.* Almqvist & Wiksell, Stockholm.

Ingram, H. A. P. (1983) Hydrology, Ch. 3, pp. 67–158 in Gore, A. J. P. (ed.) *Ecosystems of the World 4A. Mires: Swamp, Bog, Fen and Moor.* Elsevier, Amsterdam.

Iversen, J. (1949) The influence of prehistoric man on vegetation. *Danm. geol. Unders.* **RIV(3)** no. 6.

Kershaw, K. A. (1973) *Quantitative and Dynamic Plant Ecology*, 2nd edn. Edward Arnold, London.

Klötzli, F. (1970) Eichen-, Edellaub- und Bruchwalder der Britischen Inseln, *Schweiz, Z. Forst.* **121**, 329–66.

Kulczynski, S. (1949) Peat bogs of Polesie, *Mem. Acad. Polon. Sci. Lett. Cl. Sci. Math. Nat., Ser. B Sci. Nat.* **15**, 1–356.

Kunick, W. (1982) Comparison of the flora of some cities of the central European lowlands, in Bornkamm, R., Lee, J. & Seaward, M. (eds) *Urban Ecology.* Blackwell Scientific Publications, Oxford.

Larcher, W. (1975) *Physiological Plant Ecology.* Springer-Verlag, Berlin.

Lohmeyer, W. *et al.* (1962) Contribution a l'unification du systéme phytosociologique pour l'Europe moyenne et nordoccidentale, *Melhoramento* **15**, 137–51.

Madgwick, H. A. I. & Ovington, J. D. (1959) The chemical composition of precipitation in adjacent forest and open plots, *Forestry, 32*, 14–22.

McVean, D. N. & Ratcliffe, D. A. (1962) *Plant Communities of the Scottish Highlands.* HMSO, London.

Maitland, P. S. (1978) *Biology of Fresh Waters.* Blackie, Glasgow.

Matthews, J. R. (1955) *Origin and Distribution of the British Flora.* Hutchinson, London.

Moore, J. J. (1962) The Braun-Blanquet system: a re-assessment, *J. Ecol.* **50**, 761–70.

Moore, J. J. (1968) A classification of the bogs and wet heaths of northern Europe (Oxycocco-Sphagnetea Br.-Bl. & Tx. 1943), pp. 306–20 in Tuxen, R. (ed.) *Pflanzensoziologische Systematik*, Den Haag.

Moore, J. J., Fitzsimons, P., Lambe, E. & White, J. (1970) A comparison and evaluation of some phytosociological techniques, *Vegetatio* **20**, 1–20.

Moore, P. D. & Bellamy, D. J. (1973) *Peatlands.* Elek Science, London.

Moore, P. D. & Webb, J. A. (1978) *An Illustrated Guide to Pollen Analysis.* Hodder and Stoughton, London.

Nature Conservancy Council (1984) *Nature Conservation in Great Britain.* Nature Conservancy Council, Peterborough.

Odum, E. P. (1971) *Fundamentals of Ecology*, 3rd edn. Saunders, Philadelphia.

Pears, N. (1977) *Basic Biogeography.* Longman, London.

Pearsall, W. H. (1950) *Mountains and Moorlands.* Collins, London.

Pennington, W. (1969) *The History of British Vegetation.* English Universities Press, London.

Perring, F. P. & Walters, S. M. (eds) (1962) *Atlas of the British Flora.* Nelson, London.

Peterken, G. (1981) *Woodland Conservation and Management.* Chapman and Hall, London.

Polunin, N. (1960). *Introduction to Plant Geography.* Longman, London.

Poore, M. E. D. (1955) The use of phytosociological methods in ecological investigations. I. The Braun-Blanquet system, *J. Ecol.* **43**, 226–44.

Rackham, O. (1976) *Trees and Woodland in the British Landscape.* J. M. Dent and Sons, London.

Rackham, O. (1986) *The History of the Countryside.* J. M. Dent and Sons, London.

Rainfall Atlas of the British Isles (1926). Royal Meteorological Society, London.

Ranwell, D. S. (1972) *Ecology of Salt Marshes and Sand Dunes.* Chapman and Hall, London.

Ratcliffe, D. A. (1977) *A Nature Conservation Review*, vol. 1. Cambridge University Press, Cambridge.

Ratcliffe, D. A. (1984) Post-Medieval and recent changes in British vegetation: the culmination of human influence, *New Phytol.* **98**, 73–100.

Raven, J. & Walters, M. (1956) *Mountain Flowers.* Collins, London.

Richards, P. W. (1952) *The Tropical Rain Forest.* Cambridge University Press, Cambridge.

Rieley, J. O., Page, S. E. & Shah, A. A. (1984) Eutrophication of afforested basin mires in the midlands of England, *Proc. 7th Int. Peat Cong., Dublin* **1**, 375–87.

Rieley, J. O., Richards, P. W. & Bebbington, A. D. L. (1979) The ecological role of bryophytes in a North Wales Woodland. *J. Ecol.* **67**, 497–527.

Salisbury, E. (1952) *Downs and Dunes.* G. Bell and Sons, London.

Schmid, J. A. (1975) *Urban vegetation. A review and Chicago case study.* University of Chicago, Department of Geography research paper no. 161, Chicago.

Sheail, J. (1971) *Rabbits and their History.* David and Charles, Newton Abbot.

Shimwell, D. W. (1971) *Description and Classification of Vegetation.* Sidgwick and Jackson, London.

Smith, A. J. E. (ed.) (1982) *Bryophyte Ecology.* Chapman and Hall, London & New York.

Street, H. E. & Opik, H. (1984) *The Physiology of Flowering Plants*, 2nd edn. Edward Arnold, London.

Sukopp, H., Blume, H. P. & Kunick, W. (1979) The soil, flora and vegetation of Berlin's waste lands, pp. 115–32 in Laurie, I.C. (ed.) *Nature in Cities.* Wiley, Chichester.

Sukopp, H. & Werner, P. (1982) *Nature in Cities.* European Committee for the Conservation of Nature and Natural Resources, Strasbourg.

Sutcliffe, J. F. & Baker, D. A. (1974) *Plants and Mineral Salts.* Edward Arnold, London.

Szafer, W. (1966) *The Vegetation of Poland.* Pergamon Press, Oxford.

Tallis, J. H. (1983) Changes in Wetland Communities, Ch. 9, pp. 311–47 in Gore, A. J. P. (ed.) *Ecosystems of the World 4A. Mires: Swamp, Bog, Fen and Moor.* Elsevier, Amsterdam.

Tansley, A. G. (1939) *The British Islands and their Vegetation.* Cambridge University Press, London. (Reprinted 1949.)

Taylor, J. A. (1983) The Peatlands of Great Britain and Ireland, Ch. 1, pp. 1–46, in Gore, A. J. P. (ed.) *Ecosystems of the World 4B. Mires: Swamp, Bog, Fen and Moor.* Elsevier, Amsterdam.

Townsend, W. N. (1973) *An Introduction to the Scientific Study of the Soil,* 5th edn. Edward Arnold, London.

Troels-Smith, J. (1964) The influence of prehistoric man on vegetation in central and north-western Europe, *Proc. VI Congr. INQUA* **2**, 487.

Vince-Prue, D. (1975) *Photoperiodism in Plants.* McGraw-Hill,.

Walter, H. (1973) *Vegetation of the Earth.* English Universities Press, London.

Walters, S. M. (1970) The next twenty-five years, pp. 136–41 in Perring, F. P. (ed.) *The Flora of a Changing Britain.* E. W. Classey Ltd, Hampton, Middlesex.

Westhoff, V. & Den Held, A. J. (1969) *Plantengemeenschappen in Nederland.* Thieme, Zutphen.

Wheeler, B. D. (1980) Plant communities of rich-fen systems in England and Wales. II. Communities of calcareous mires, *J. Ecol.* **68**, 405–21.

Whittaker, R. H. (1970) *Communities and Ecosystems.* Macmillan, New York. (2nd edn. 1975.)

Whittaker, R. H. (1973) Approaches to classifying vegetation, in Whittaker, R. H. (ed.) *Vegetation Science: Part V. Ordination and Classification of Communities.* Junk, The Hague.

Species index

Index to phytosociological nomenclature

Subject index